Tao Science

Tao Science

The Science, Wisdom, and Practice of Creation and Grand Unification

DR. AND MASTER
Zhi Gang Sha
&
Dr. Rulin Xiu

The information contained in this book is intended to be educational and not for diagnosis, prescription, or treatment of any health disorder whatsoever. The content of this book is intended to be used as an adjunct to a rational and responsible healthcare program prescribed by a healthcare practitioner. The author and publisher are in no way liable for any misuse of the material.

Published by Waterside Press and Heaven's Library Publication Corp.

Waterside Press
2055 Oxford Ave.
Cardiff, CA 92007
www.waterside.com

Heaven's Library Publication Corp.
30 Wertheim Court, Unit 27D
Richmond Hill, ON L4B 1B9
heavenslibrary@drsha.com

ISBN 13: 978-1-939116-22-2 print on demand edition
ISBN-13: 978-1-947637-68-9 print edition
ISBN-13: 978-1-947637-69-6 ebook edition

Design: Lynda Chaplin

Contents

Introduction

Our Journey to Tao Science

OUR JOURNEY TO TAO SCIENCE is the meeting of two souls that followed completely different life paths. Dr. and Master Zhi Gang Sha followed the path of service to serve all humanity. The path of service is the highest spiritual practice. Through Dr. and Master Sha's tireless and selfless service with his whole heart, the Creator has bestowed upon Master Sha the deep wisdom, profound knowledge, effective practices, and miraculous powers to help humanity evolve to a higher level. This wisdom, knowledge, and practices are simple and powerful. They have healed, transformed, and uplifted the lives of millions of people. Master Sha has also been teaching and leading thousands of students on the journey to reach Tao. Tao is the ultimate Source. To reach Tao is to reach a state of complete freedom, love, power, bliss, and enlightenment, including moving toward immortality.

On this journey, a student named Dr. Rulin Xiu appeared. A quantum physicist and string theorist, Dr. Xiu was touched and inspired by Master Sha's teaching, wisdom, knowledge, and practices, and especially by his selfless and tireless service to humanity. She was empowered by Master Sha's teaching, healing, and blessing to help integrate Tao wisdom into science. Together, they have created Tao Science and are honored to introduce it in this book. Dr. Xiu's experience of working on Tao Science illustrates what is possible for everyone when one applies the wisdom and practice of Tao, the ultimate Source. She is delighted to share her journey with you.

Dr. Rulin Xiu's Journey

I grew up in Xi'an, one of the oldest and greatest cities in China, and now the capital of Shaanxi province. Many people know the city by the life-sized terracotta army from the first Emperor of China's[1] tomb.

As a child, I showed some talent in mathematics and science. After I graduated from high school, I was recruited to the University of Science and Technology of China (USTC) in Hefei, Anhui province. At the time, USTC was the leading university in China for science and technology. It is one of the top nine universities in China. All USTC students are at the top of their high school classes. The academic atmosphere of USTC was quite competitive. USTC was the training ground for the Chinese Academy of Sciences, the top scientific research institute in China.

In my first year of college, a friend gave me a small booklet called *Physics and Simplicity* by the American quantum physicist, John Wheeler. This small book completely changed my life.

John Wheeler was the last living physicist who worked with both Nobel Laureates Albert Einstein, the founder of Relativity Theory, and Niels Bohr, one of the important founders of quantum physics. For many years, Wheeler and his students worked on reconciling quantum physics with relativity in order to build a unified physics theory that includes both. Wheeler is also known for popularizing the term "black hole," for coining the terms "neutron moderator," "quantum foam," "wormhole," and "it from bit," and for hypothesizing the "one-electron universe."

---

[1] Qin Shi Huang (秦始皇), 259-210 BCE

John Wheeler had given a series of lectures at USTC, a couple of years before I matriculated in 1982. His small book, *Physics and Simplicity*, was a compilation of those lectures. One of the ideas presented in the book inspired me deeply. The idea is to develop one theory to explain and understand everyone and everything. This idea is the dream of some people. Many physicists call it the Grand Unification (or Unified) Theory (GUT). In physics, GUT is to use one mathematical formula to account for everything in the cosmos. It is the holy grail of modern physics. The outstanding physicists who have worked on this problem include Isaac Newton and Albert Einstein.

At the age of eighteen, without telling anyone, I knew deep within myself that developing the GUT is one of my tasks in this lifetime. I have no choice but to accomplish this task.

John Wheeler's book also inspired my interest in quantum physics and Relativity Theory. In fact, I became completely obsessed with them. I ignored my classes, spending most of my time reading papers and books by the founders of quantum physics. A friend gave me a book of the complete collection of Einstein's scientific papers. I read the book several times.

I also started to read philosophy and even psychology books. Intuitively, I knew that a complete understanding of quantum physics had to include our consciousness and a deeper understanding of our very existence. However, the more books I read, the more confused I became. I was not alone in my confusion in trying to understand quantum physics. One of my favorite quantum physicists, Nobel Laureate Richard Feynman, once said, "I think I can safely say that nobody understands quantum mechanics." Another Nobel Laureate in Physics, Steven Weinberg, candidly admitted, "There is now in my opinion no entirely satisfactory interpretation of quantum mechanics."

I spent many of my college days reading *The Feynman Lectures on Physics*. However, I didn't have the good fortune to meet Prof. Feynman. When I entered University of California at Berkeley for graduate studies in 1989, he had already passed away (February 15, 1988) due to two rare forms of cancer.

I did get the chance to meet Steven Weinberg. He attended my seminar in the Theory Group of Physics Department at the University of Texas in 1995 and made some astute comments about my research on string phenomenology and Grand Unification Theory. Steven Weinberg contributed seminal work on the unification of weak and electromagnetic interactions and predicted the existence of the Higgs boson, popularly known as the "God particle." The appearance of this particle provides a way to explain how other particles get their masses.

Frustrated with not being able to reach a good understanding of quantum physics, I turned my focus to solving mathematical problems in physics, like most physicists. My graduate research at Berkeley was to study string theory as GUT. String theory is a branch of quantum physics that studies how the movement of a string could create all of the particles and forces found in nature.

In 1994, I received my Ph.D. in Physics with relative ease. My dissertation was entitled, "Grand Unification Theory and String Theory." I then continued my research on GUT at the Houston Advanced Research Center, but no matter how hard I tried, I couldn't make the progress I really wanted on GUT. I was not alone in this blockage either. Many more brilliant theoretical physicists than I have had the same experience. Steve Weinberg expressed the situation well: "The more the universe seems comprehensible, the more it also seems pointless."

In 1996, in an effort to help a friend with his business in China, I found myself starting to do business. As a result, I completely forgot about my dream to find GUT until I met Master Sha in 2009, some thirteen years after I stopped my academic pursuits.

In 2003, I moved to Hawaii to set up a factory for my new business. The deep spiritual tradition in Hawaii opened me to the miraculous spiritual realm. During a spiritual celebration, I had a profound encounter with the Divine. The Divine had me experience and realize the most profound secret about divine love: *Everyone and everything in the world is the Divine's love.* There is only love. Love is the only truth. Love is the only existence. The experience of divine love changed my life completely. My life became bliss and a stream of miracles. Almost every thought I had seemed to manifest almost instantly. I was also led to meet my spiritual teacher, Dr. and Master Sha.

On September 9, 2009, I was urged by a female kahuna to attend a workshop given by Master Sha in my neighborhood on the Big Island of Hawaii. A kahuna is a revered medium and priest in Hawaii who can receives messages from the spiritual realm in order to help with human affairs. In the past, high-level kahunas have been trusted counselors for kings and queens on their most important decisions.

Master Sha impressed me right from the beginning. He started the workshop without knowing what he was going to talk about. He checked and told us that he received the message that he should teach us Tao that day.

Learning about Tao has been one of my passions. In college, I already sensed that Tao could provide the key to GUT. When I came to the United States for my graduate study, I brought with me many books about Tao. I read them almost daily. However, I never could figure out how to apply Tao wisdom to GUT.

The way Master Sha taught Tao amazed me. I have never heard such profound knowledge explained in such simple and clear language. Master Sha's healing power was also truly astounding. If I had not seen it first hand, I would not have believed it. But, what really touched my heart was his complete commitment to serve others and humanity.

After the workshop, as Master Sha signed his books for me, I told him I wanted to become his student. Little did I know at that moment that Master Sha's wisdom, knowledge, and healing work held the key to unlock GUT. Becoming his student is one of the most important steps I have ever taken on my journey towards finding a satisfactory solution to GUT. After that fateful day, I started to attend almost every one of Master Sha's workshops to learn about the soul, soul healing, and Tao.

One of the spiritual abilities Master Sha trains students to develop is to open our spiritual communication channels to receive messages, images, sounds, words, and other information from the cosmos. These spiritual channels include the Soul Language Channel, the Direct Soul Communication Channel, the Third Eye Channel, and the Direct Knowing Channel. Master Sha's ability to receive information is quite astonishing. His teachings and books all come from "flow," meaning he connects with the Source and lets the information flow through his speech. During some workshops, he has flowed parts of his books. It is extraordinary to witness. Heaven literally opens, showering treasures, nectar, jewels, pearls, and all kinds of wisdom on us. To be in his presence is beyond words.

Master Sha mentioned quite a few times in his workshops that quantum physics is the physics about soul. I would just sit there trying to understand this statement. Finally, in one of these workshops, I had an "aha!" moment. I suddenly realized that it is possible to use quantum physics to mathematically describe soul. Soul is not as mysterious and beyond this physical world as most of us normally think! It can be defined as a physical quantity, similar to mass, weight, force, and charge.

I told Master Sha about this realization. He saw its significance immediately. And so, we initiated the creation of the Soul Mind Body Science System[2], which in turn paved the way to Tao Science. The process of creating the Soul Mind Body Science System and Tao Science has been a beautiful journey. It would take another book or two to lead you through this spectacular cosmic pilgrimage. I will just mention a few key points here.

First, Master Sha spiritually created a Soul Mind Body Science System Committee and a Tao Science Committee. Together, these committees include one hundred and eleven known and unknown saints and scientists in Heaven and historically on Mother Earth.

Second, Master Sha spiritually removed all kinds of blockages that have prevented such a science from being established. This process included karma cleansing for science and for me. We will speak much more about karma in this book, particularly in chapter ten.

Third, Master Sha transmitted the wisdom, knowledge, and practical techniques of the Soul Mind Body Science System and Tao Science in spiritual form to himself and me.

---

[2] Dr. and Master Zhi Gang Sha & Dr. Rulin Xiu: *Soul Mind Body Science System: Grand Unification Theory and Practice for Healing, Rejuvenation, Longevity, and Immortality*. Dallas/Toronto: BenBella Books/Heaven's Library Publication Corp., 2014.

Fourth, Master Sha wrote one-stroke (*yi bi zi,* 一笔字) Tao Calligraphies with the phrases, "Soul Mind Body Science System" and "Tao Science" in Chinese. He transmitted billions of saints and saint animals from the Tao realms to these calligraphies. We will share more about the power and significance of Tao Calligraphy in chapter nine.

This spiritual work has been essential for creating the Soul Mind Body Science System and Tao Science. Every day, I meditate in front of the calligraphies and ask the Divine, Tao, Soul Mind Body Science Committee, Tao Science Committee, saints, and saint animals to guide me in our creation of Soul Mind Body Science System and Tao Science. Lao Zi, the Buddha, Mayan saints, Richard Feynman, Isaac Newton, Albert Einstein, and many others have appeared to us in spiritual form to help us. Without their help and assistance, we would never have been able to receive such a beautiful and powerful scientific framework to unify physical, spiritual, and conscious existence as one. Without their help, we would not have been able to see the mathematical expression of Tao Science and write this book. We are so grateful for all the help we have received for creating the Soul Mind Body Science System and Tao Science.

One day in January 2013, I was talking with Master Sha about my dream of coming up with the solution to GUT. Master Sha closed his eyes. This is the way he usually connects with Tao, the ultimate Source. When he opened his eyes, he picked up a pen and wrote the formula:

$$\mathbf{S + E + M = 1}$$

He said, "This is the grand unification formula."

I looked at the formula, astonished at its simplicity but totally lacking in comprehension. I asked Master Sha, "What are S, E, and M in the formula?"

He told me, "S is *shen* (神), which includes soul, heart, and mind. E is *energy*. M is *matter*. The number '1' represents the grand unified field." This formula contains everyone and everything. It contains all wisdom and practices. It contains all laws and principles. It contains all healing and rejuvenation. It contains transformation for every aspect of life.

For the last several years, I have been trying to understand this formula fully. At times, I feel I have started to really understand it. There are even moments when I think I have come to a complete comprehension. But, the next moment or the next day, I realize I know almost nothing about it.

I am so humbled by and grateful for this receiving process. Truly, Tao Science, the Grand Unification Theory we have received from the Divine and Tao, is more powerful, beautiful, profound, and satisfying than anything I could have ever imagined or dreamed. This science, wisdom, knowledge, and practice are given to humanity now because they are urgently needed to save humanity and Mother Earth.

On August 6, 2014, I was on the phone with Master Sha and Master Sha's assistant, Master Cynthia, doing the final read-through of the second book of this series, *Soul Mind Body Science System: Grand Unification Theory and Practice for Healing, Rejuvenation, Longevity, and Immortality*. Sitting on the balcony of my oceanfront home on the Big Island of Hawaii, I noticed and made a comment that the ocean waves were getting unusually close to my house, less than thirty feet away. Through their Third Eyes, both Master Sha and Master Cynthia saw the spiritual image that my house would be swept away by the ocean waves. They urged me to leave the house immediately with only the most necessary belongings.

I took Master Sha's Tao Calligraphy scrolls down from the wall, gathered some clothes and my laptop, and called my two dogs over to get into the car. But, the ocean waves had already blocked the road from the garage. Since it was not safe to drive, I walked out of my property with my two dogs.

I stayed one night at my friends' home with the heavy storm howling. The next day in calmer weather, my friends drove me to what had been my home. We were shocked to discover that my house was the only house in all of the Hawaiian Islands to be completely devastated by the storm.

Waves had rushed in and knocked down the whole house. The entire roof lay on the ground and most of my belongings had washed away. My car, refrigerator, and washing machine were washed hundreds of yards away, finally to be stuck between coconut trees. Walking around my property with my friends, I was so shocked by the force of nature that I could not talk, think, or do anything.

Finally, sitting in the back of the car as we drove away, I cried. I didn't cry for the loss of my house and belongings. I had often thought about tearing down the house and building my dream home in its place. Now, the Divine did the job for me. I cried because I actually knew that this "disaster" was the Divine's highest love for me. It was the Divine's urgent calling to me to go out to spread divine love, the Soul Mind Body Science System, and Tao Science.

I had received this message from the Divine before. Once during meditation, the Divine revealed to me the horrendous disasters coming to humanity. The Divine told me, "If you don't take action, this is what will be happening to humanity and Mother Earth." At that moment, I was shocked. I didn't know I could matter that much.

Seeing what happened to my house, I finally realized the seriousness of the message and how urgent it was in the Divine's heart. And so, I was moved to tears. I could hear the Divine's loving voice softly speaking to me:

"My beloved daughter, Rulin, it is time for you to step up to be my messenger to awaken humanity to my love. It's time for you to awaken humanity to their souls, to the infinite potential within each human being. It's time for you to travel and spread the wisdom, knowledge, and practice of divine love, the Soul Mind Body Science System, and Tao Science so that humanity can be uplifted to a higher level."

In the Divine's love and urging, I started a new chapter of my life. My life has transformed greatly. In May 2015, I was appointed by Master Sha to become one of his disciples and Worldwide Representatives, and a Divine Channel. It is truly the highest honor in all my lifetimes. The upliftment I felt with these appointments is beyond any words or comprehension.

This greatest upliftment comes with the greatest abilities, as well as the greatest responsibilities and purification. I realize that every word I speak, every thought I think, every sight I see, every sound I hear, every feeling I have, every movement I make, every flavor I taste, and every fragrance I smell holds significance not just for myself, but also for humanity, for Mother Earth, and for countless souls in the whole universe and all universes. To help me grow further, Master Sha also gave me serious teaching about how my ego could block me from growing and receiving the purest messages from the Divine. He wrote a Tao Calligraphy *Tao Qian Bei* (道谦卑, *Tao humility*), to help me remove my ego.

I flew back to Hawaii and stayed on my property for a few weeks to purify myself. I received the Divine's message that this period of purification was also for receiving a physics paper. My home at that moment consisted of a little shack my friends put up for me. I had no running water, no working toilet, no electricity. Rats and ants consider this their official home now for a good reason. This was truly the best place for me to purify and connect with the Divine.

My friends had put great love into putting the place together. To my greatest astonishment, John Wheeler's little book, *Physics and Simplicity*, mysteriously and miraculously sat on the bookshelf. When I came to United States, this was one of the few western books I brought with me from China. When I moved to Hawaii, I had already stopped doing physics academically. This was the only physics book I took to Hawaii. I had thousands of books in my library in my Hawaii beach home. At that moment, on the broken bookshelf, stood only five surviving books, including *Physics and Simplicity* and Lao Zi's *Dao De Jing*.

I hadn't read *Physics and Simplicity* for more than twenty years. The mysterious appearance of the book reminded me of the significant role John Wheeler played in my life. I looked deeper into his work. I was astonished to find that Wheeler's "it from bit" summarized half of the basic principles of Tao Science. I want to express my greatest gratitude to John Wheeler for his inspiration and contribution to my life and to physics. I hope this book and my work with Master Sha can help show my deepest gratitude, appreciation, and acknowledgement of this important physicist.

The beautiful Hawaiian ocean, mountains, plants, animals, gods, and goddesses are among the most nourishing and loving in the world. The purification and bliss I felt in my soul, heart, mind, and body were beyond words. An original physics paper flew out of me like intoxicating and exhilarating hot spring water in two days. The paper[3] is about how to derive a formula to describe our universe and how to calculate the dark energy, dark matter, and vacuum energy, which is called the cosmological constant, from this formula.

To calculate the cosmological constant from fundamental physics theory has been one of the most challenging and important problems in theoretical physics. The solution presented in the paper is surprisingly simple yet powerful. I was astounded by the beauty and simplicity of the paper.

One night, the idea came to me to present this paper at a physics conference. Searching the internet, I found the 4th International Conference on New Frontiers in Physics. Truly, this would be the right place to present. Unfortunately, the application deadline had already passed. However, feeling I had nothing to lose, I emailed an abstract of the paper to the organizers and then went to sleep. When I woke up the next morning, to my greatest amazement, I saw on my phone an email invitation to attend and present at the conference.

On August 25, 2015, I flew to Crete to present our physics research. My intention was to show the physics community that Tao Science could solve some challenging problems in string theory and GUT. Considering that most physicists are not ready to hear about the spiritual aspects of our work, I decided not to mention Tao, spirituality, soul, heart, mind, or the Soul Mind Body Science System at all. But, I spilled out the secret when an attendee asked me at the end of my presentation what other results we could obtain through our

---

[3] Dr. Zhi Gang Sha and Dr. Rulin Xiu. "Dark Energy and Estimate of Cosmological Constant from String Theory" to appear in *Journal of Astrophysics & Aerospace Technology* (accepted on March 27, 2017).

approach. I said that this work is part of a larger project that intends to bring science and spirituality together.

After my presentation, the next presenter spoke about Albert Einstein's work. During this session, Einstein appeared to me in spiritual form. He was full of sorrow. I realized the Divine sent Einstein to deliver a message to me about the social responsibility of theoretical physicists.

As soon as we had a break, I told one of the organizing chairs of the conference about our work on how to integrate science and spirituality as one at the fundamental level. The organizers of the conference also considered this to be an important subject. They asked me to email them an abstract of our work. The next day, to my greatest delight, they told me that they found a time slot for me to present the Soul Mind Body Science System at the conference.

I could hear the Divine talking to me, telling me what I needed to say. I wrote down those words and tried to memorize them for my presentation. The following is part of the message I had written down and delivered at the conference:

"I am very grateful to the organizers of this conference to give me the opportunity to share our research about unifying science and spirituality.

"In recent years, this subject of unifying science with spirituality has been getting more and more public attention. I believe it is an extremely important task for physicists at this moment.

"Since the scientific revolution that consummated in Newton's three laws, physical science and spirituality have become separated. This separation has caused great disharmony within ourselves, in our lives, in our societies, in humanity, and in our world. I believe this separation of science and spirituality is one of the major causes for the escalating environmental, financial, social, spiritual, emotional, mental, and physical challenges facing humanity. I believe that, as physicists, we carry a major responsibility right now to bring physical science and spirituality back together. This is crucial to help humanity pass through the upcoming difficult times."

On October 19, 2015, I gave a presentation at the Parliament of the World's Religions. This is one of the largest interfaith movement gatherings in the world with nearly ten thousand attendees from more than eighty religions and fifty nations. On this occasion, we once again called for the unification of science and spirituality:

"The unification of science and spirituality is urgently needed. It will help save humanity from more suffering. It will take both science and spirituality to new heights. Our dream is that science and spirituality can join as one to enrich each other and uplift each other.

"Now more than ever, we need the Grand Unification Theory and Practice not only to unify the fundamental physical forces, but also to unify our souls, hearts, minds, and bodies; to unify science with love and spirituality; and to unify humanity with nature, so we can live harmoniously with ourselves, each other, and our environment. More importantly, now is the time to transcend our human limitations, to connect and uplift each individual and humanity to the Divine, to our highest human potential and life purpose with more love, peace, harmony, joy, health, wisdom, beauty, abundance, and enlightenment in our lives."

Dr. Rulin Xiu
January 5, 2017

Many of you have heard the calling for the unification of science and spirituality.

Many of you are looking for the power of love to transform every aspect of your lives and the lives of others.

Many of you long to become enlightened to live in complete love, bliss, wisdom, abundance, and freedom.

Many of you are searching for the elixir of immortality, the way to reach eternal life.

We are honored to be on this journey with you. We are grateful to have the opportunity to serve and share what we have received with you. We are humbled to work together with you to bring grand unification to our lives, to humanity, and to all beings through Tao Science.

List of Figures

1

What Is Tao Science?

TAO (道) IS THE SOURCE of everyone and everything. Tao Science is the science of the Source and creation. It is a science to enlighten us with knowledge about the Source and the way of the Source. It is a science that tells us what everyone and everything are made of, how everyone and everything are created, and how one can become a creator and manifestor. Tao Science is a science of grand unification. It unifies science with spirituality at the most fundamental level. It unifies everyone, everything, and every aspect of our lives.

Secrets of Creation

Understanding the secrets of creation is some of the highest wisdom, knowledge, and enlightenment that a human being can dream about. This wisdom and knowledge will not only empower us to become a powerful manifestor; more important, they are essential for liberating humanity from all kinds of illusion, suffering, limitation, lack, ignorance, and bondage. This wisdom of creation is the door to higher levels of consciousness and enlightenment.

What is the source of the universe, including everyone and everything? How are everyone, everything, and the universe created? How will they evolve? What is their final destiny? How is every aspect of our lives created? Many have asked these fundamental and important questions for centuries. The answers to these questions can help determine our lives, our society, and our world in the most profound and significant ways.

The search for the answers to these questions has become increasingly urgent as we and our planet are facing more and more challenges. What is the root cause of global warming, natural disasters, war, poverty, violence, financial upheavals, and energy crises? What is the root cause of the challenges in our relationships, finances, and health—physical, emotional, and mental?

To know the root cause of our challenges is to empower us to overcome them. It is vital for us to understand how we are created and how our realities are manifested. This understanding is crucial for humanity to resolve energy and financial crises, global warming and other environmental problems, wars, poverty, escalating health issues, depression, and anxiety, as well as many other difficulties in our lives and the world.

Creation myths exist in every culture. Much wisdom and knowledge about creation has been explored in all kinds of philosophies, ideologies, sciences, and religions. Some of their conclusions and beliefs agree. Some may not agree.

Many people have tried to find the laws and truth of creation through science. Science has increased our knowledge and power to create significantly. However, scientific answers to the fundamental questions about creation have been limited.

Physics is the foundation of the natural sciences. Physics studies the nature and behavior of matter and energy in the universe, and quantifies the movement of matter and energy through space and time with mathematical formulas. Current physics theory has some significant shortcomings. It does not tell us the source or origin of everyone and everything, how everyone and everything are created, or what the ultimate destiny is for everyone and everything. In addition, physics deals with only the physical realm. It does not address consciousness or spirituality. It cannot tell us the meaning and purpose of life.

With the power of science to create increasing exponentially, it becomes urgent for physics to overcome its limitations. If science cannot address the fundamental questions about our existence, scientific development could cause more harm to humanity. Science without heart and soul can be extremely dangerous. History speaks for itself. With the accelerating development of science in the last few centuries, we have experienced two unprecedented world wars. Industrialization has caused irreversible damage to our environment. Powerful weapons that can wipe out much of humanity are stockpiled by many governments and other groups.

Medical science has also progressed rapidly. However, its limitations are also becoming more and more obvious. The number of sicknesses grows faster than methods of treatment can be developed. The cost of medical treatment—where it is available—is becoming too high for most people and societies to bear. Some treatments can create new harm along with any benefits. We honor modern medicine, but it clearly has its limitations.

Many people turn to spirituality and religion for answers about creation. A few sages, gurus, saints, buddhas, holy beings, and other spiritual masters have obtained deep and profound secrets and wisdom about creation. They received the secrets and wisdom through direct contact and experience with the Source and Creator.

However, although these sages and saints can express the secrets and wisdom about creation in simple words, they are still beyond the grasp of most people. This is because these profound truths can be obtained only at a high level of consciousness. A similarly high level of consciousness is needed to understand and realize these truths. It generally takes many lifetimes of dedicated spiritual practice to reach such a high level of consciousness. Consequently, it is out of reach for most people.

Understanding the laws and truth of creation is critical for reaching higher spiritual levels. Gautama Buddha, considered to be the founder of Buddhism, reached enlightenment because he obtained a profound understanding about the workings of the universe. After he attained this realization, he started to teach others and help them reach enlightenment.

Can we express the secrets and wisdom about creation scientifically? If we can, it would help science progress to an unprecedented level. It could also help more people understand the profound truths about creation and reach enlightenment.

Some may be concerned that the secrets about creation could fall into the wrong hands and be misused. There is no need for such concern. You will learn in this book that the deepest wisdom and highest power is the greatest kindness. The greatest kindness is the highest power and deepest wisdom. Learning the laws of creation will help all beings heal, transform, and uplift to a higher level of existence.

Wisdom of Tao

The secrets and wisdom of Tao are ancient Chinese wisdom, philosophy, science, tradition, ethic, and practice. They have a written history of more than five thousand years. These secrets and wisdom are some of the most profound, sacred, and enduring truths throughout the recorded history of Chinese culture. Now, more and more people worldwide are finding and learning this wisdom.

Tao is the ultimate Source.

Tao is the universal laws and principles of the Source. Follow Tao, prosper. Go against Tao, finish. This profound truth has been known for thousands of years.

Tao is a universal practice, the Way of all life. Integrate Tao practice into daily life to have good health, harmonious relationships, and a flourishing life with more power, freedom, longevity, and even immortality.

Lao Zi is one of the greatest sages humanity has known. More than two thousand five hundred years ago, Lao Zi gave us *Dao De Jing* to teach us what Tao is, the way of Tao, and how to reach Tao. In the first lines of *Dao De Jing*, Lao Zi tells us:

> Dao ke Dao
> Fei chang Dao

which means *the Tao that can be spoken is not the ultimate Tao.*

In 2009, Master Sha received Tao Jing, the Classic of Tao, from the Source. This sacred and secret wisdom further reveals:

> Da wu wai
> Xiao wu nei
> Wu fang yuan
> Wu xing xiang
> Wu shi kong

Shun Dao chang
Ni Dao wang

which means:

(Tao is) Bigger than biggest
Smaller than smallest
No form
No shape or image
No time or space
Follow Tao, flourish
Go against Tao, finish

How is everything created from Tao? Lao Zi teaches us in *Dao De Jing* that everything is created through the following process:

Dao sheng yi, yi sheng er, er sheng san, san sheng wan wu.

Tao creates One. One creates Two. Two creates Three. Three creates everyone and everything.

This is Tao Normal Creation.

What is the destiny of everyone and everything? Master Sha received the wisdom that the destiny of everyone and everything is to return to Tao through Tao Reverse Creation:

Wan wu gui san, san gui er, er gui yi, yi gui Dao.

Everyone and everything return to Three. Three returns to Two. Two returns to One. One returns to Tao.

Can we understand and formulate Tao Normal Creation and Tao Reverse Creation scientifically? Yes! This book will explain how.

Significance of Tao Science

Tao Science is based on ancient Tao wisdom combined with the mathematical framework developed in quantum physics. Tao Science expresses Tao wisdom through mathematical formulas. Tao Science is important for each one of us and for all of humanity in three ways:

1. Tao Science brings revolutionary and cutting-edge breakthroughs in science and technology.

   In our research, we have found that all current physics can be derived from Tao Science. More important, Tao Science sheds light on and helps us solve a number of challenging problems and questions in quantum physics, string theory, astrophysics, cosmology, and the Grand Unification Theory, as well as in the science of consciousness, philosophies, and more.

2. Tao Science provides a scientific way to understand profound spiritual wisdom. It can help us know and realize our highest power, greatest potential, and deepest meaning of our lives. It can help all humanity evolve to higher levels of existence and enlightenment.

3. Tao Science can help save each of us, all humanity, and Mother Earth from more suffering and conflict. It can bring grand unification within each of us and with all beings.

Do you want to create the life you want? Would you like to become a powerful manifestor? Would you like to know and develop your higher powers, greater abilities, and deeper potentials? Would you like to know how to transform your relationships, finances, intelligence, and every aspect of your life?

Would you like to help others have a better life? Would you like to help the world become a better place? Would you like to have greater meaning for your life? Would you like to reach enlightenment and immortality?

If your answer to any of these questions is *yes*, please read on.

2

Milestones of Physics

PHYSICS IS THE NATURAL SCIENCE that studies matter, its constitution, and its motion through space and time. Physics often provides the explanation of the basic mechanisms of other natural sciences. It also opens new avenues of research by giving new concepts, new understanding, and new analytical tools, including new experimental devices and equipment. Because of this, physics is often called the foundation or root of natural science.

Origins of Physics

The word *physics* came from the ancient Greek word meaning "knowledge of nature." The earliest recorded physics research was in the form of what we now classify as astronomy and astrology. Astronomy studies heavenly bodies, such as planets, stars, and other objects in outer space. Astrology studies the influences of the positions of planets, stars, and other heavenly bodies on human affairs and the events on Mother Earth.

Early civilizations dating back to before 3000 BCE, including ancient Sumerians, Chinese, Indians, Egyptians, and many others, all had impressive understanding of the motions of the sun, moon, stars, and planets. Some ancient civilizations worshiped planets and stars as their gods. Ancient Chinese believed Heaven and human beings are one. What happens in Heaven directly affects humanity. Therefore, observation and knowledge about the motions of the planets and stars was important for their lives. In India, the first written record of astronomical concepts comes from the Vedas, the oldest literature of Hinduism.

Unlike other ancient civilizations, the ancient Chinese separated astronomy and astrology. The main job of the ancient Chinese astronomers was to track time, announce the first day of every month, and predict lunar eclipses. Their forecasts were vital to the emperors. If they were wrong in their predictions, they could be beheaded! Because of this, the Chinese astronomers generated fantastically accurate measurements of time and charted unusual cosmological phenomena, such as novae, comets, meteor showers, solar flares, and sunspots, long before any other culture made any such observations. This makes their work important to the historical development of astronomy. Their ideas traveled along the Silk Road into the Middle East and Europe.

The measurement of time is crucial for the development of civilization and science. Ancient civilizations used sundials to measure time. A sundial is a device that tells the time of day by the position of the sun in the sky. One of the greatest of Chinese astronomers, Guo Shoujing (1231–1316 CE), created a huge sundial that allowed him to calculate the length of a year to an accuracy of less than thirty seconds. In addition to the four major ancient Chinese inventions (compass, gunpowder, paper, printing technology), ancient Chinese also created the ancestor of the modern clock.

For over two millennia, physics has been part of natural philosophy. Natural philosophy has its origins in ancient China, Greece, India, and other civilizations.

The ancient Chinese developed some profound and advanced natural philosophies. Yin Yang, *I Ching*, Tao wisdom, and Five Elements Theory have a written history of about five thousand years. These natural philosophies explain the origin, creation, evolution, composition, and destiny of our universe, everyone, and everything as natural processes with natural causes.

In India, sacred books and hymns discuss philosophical questions about the origin of the universe. Ancient Hindu texts proposed intelligent speculations about the genesis of the universe from nonexistence.

Many ancient Greeks believed gods and goddesses controlled their lives. Some ancient Greek philosophers such as Socrates, Plato, and Aristotle rejected this non-naturalistic explanation of natural phenomena. Socrates proposed to seek truth to guide one's life and discover the truth through reason and logic in discussion. Aristotle proposed ideas that are verified by reason and observation to explain natural phenomena. Ancient Greeks recognized the holiness and wisdom of numbers along with harmony and music. They also hypothesized atomism, a theory that everyone and everything is made of atoms. This theory was found to be correct approximately two thousand years after it was first proposed.

Islamic scholars inherited natural philosophy from the Greeks. During the Islamic Golden Age (8th century–13th century CE), early forms of the scientific method were developed. Their most notable innovations were in the field of optics and vision. Optics studies the behavior and properties of light, including its interaction with matter and the construction of instruments that use or detect it. The most influential Islamic scientific books are the seven volumes of the *Book of Optics*. This treatise on optics was written by the medieval Arab scholar Ibn al-Haytham, known in the West as Alhazen (c. 965–c. 1040 CE).

The translation of the *Book of Optics* into European languages had huge impacts on the development of science in Europe between 1260 and 1650. European scholars were able to build the same devices as Islamic scholars had created more than seven hundred years earlier. From this, such important devices as eyeglasses, magnifying glasses, telescopes, and cameras were developed.

Scientific Method

Physics emerged as a unique discipline in its own right in the seventeenth century. Isaac Newton and other pioneers firmly established what we now call the scientific method. (Credit also goes to precedents in the Islamic Golden Age.) They started to use experiments and quantitative methods to discover the laws of physics. Physics has two essential elements. One is the repeatable experiment. The other is mathematics. Physicists use mathematical formulas to describe the findings of repeatable experiments.

Newton, together with other physicists, also enlightened humanity with the profound truth that planets, stars, Mother Earth, human beings, and all things follow the same laws of physics. With mathematics, the language of our logical mind, we can understand and grasp the natural laws that govern everyone and everything.

Success in explaining natural phenomena with natural laws and mathematical formulas was a monumental breakthrough for humanity. The scientific method has proven to be very powerful in expanding human knowledge about nature and our ability to create material things. Because of it, physicists are able to understand better what everything is made of at the microscopic and macrocosmic levels. The results include the creation of many of the things that are part of our daily lives, such as home appliances, TVs, DVDs, computers, the internet, airplanes, rockets, satellites, telescopes, microscopes, and advanced medical equipment, as well as nuclear weapons and much more.

Significance of Physics

Physics has deep impacts on humanity. Physics and natural science have played increasingly important roles in every aspect of our lives, our societies, and our world. Because of physics, humanity's power to create and impact nature started to grow exponentially. In approximately the last three hundred years, the whole history of humanity and Mother Earth has been transformed.

Physics is instrumental in the development of recent human history. For example, the development and establishment of Newtonian mechanics propelled and consummated the scientific revolution. Newtonian mechanics uses a set of physics laws to describe the motion of bodies under the influence of force.

The scientific revolution has greatly influenced in turn the intellectual social movement known as the Enlightenment. This began as a philosophical movement that dominated the world of ideas in Europe in the eighteenth century. The Enlightenment included a range of ideas centered on reason as the primary source of authority and legitimacy. It advocates ideals such as liberty, equality, progress, tolerance, fraternity, constitutional government, and ending the abuses of the church and state. It emphasizes scientific rigor and questions religious orthodoxy. The ideals of the Enlightenment were incorporated into the United States Declaration of Independence and the United States Constitution.

Classical thermodynamics studies movement under pressure and heat, and the exchange of matter and energy. Discoveries in thermodynamics led to the invention of the steam engine. This started the Industrial Revolution.

The founding of information theory brought humanity to the Information Age. Information theory studies the quantification, storage, transportation, and communication of information.

Physics is divided into classical physics and modern physics. Classical physics includes Newtonian mechanics, optics, thermodynamics, and electromagnetism. Modern physics includes Einstein's Special and General Theory of Relativity, quantum physics, and string theory. Let's look a little more closely into these two major subcategories of physics.

Classical Physics

The rediscovery and development of optics in seventeenth-century Europe led to more accurate measurement of the movement of heavenly bodies, especially the movement of the planets in the solar system. Mathematical relationships explaining planetary motion were discovered.

The renowned British scientist and mathematician, Isaac Newton (1642–1726 CE) proposed a gravitational force between the sun and the planets. This force was what held the planets in their positions. Newton also presented the three laws of motion.

Newton's three laws of motion tell us that everything tends to remain at rest or move in one direction at a constant pace when left undisturbed. This property is called inertia. Therefore, inertia is resistance to movement. Force is needed to make an object move in a different direction or at a faster or slower speed. Mass is the physical quantity that measures the inertia of everyone and everything. The more mass you have, the greater force is needed to move you. For every force you exert on others, the exact same force will be exerted on you.

With the introduction of gravity and the three laws of motion, Newton could derive the mathematical formulas governing the motion of the planets.

As Newton was sitting under an apple tree, a falling apple hit him on the head. This gave him an "aha!" moment. He was awakened to the universality of gravity. Gravity is not only the force that holds the sun, Earth, and other planets together in space, it is also the force that holds everyone and everything on Mother Earth. The apple falls back to the ground instead of flying away to the sky because of the gravity of Mother Earth. When we jump up, we come right back to the ground because the force of Earth's gravity pulls us back.

Gravity is a force existing between everyone and everything. This force is proportional to our mass. The more mass you have, the more gravitational force you exert on others. The gravitational force between you and me is negligible compared to other forces that affect us, because our mass is small. Earth has far greater mass.

We can all feel Earth's gravity pull. Our weight is the measure of the Earth's gravitational force upon us. When you move to a higher altitude or to outer space, you weigh less because Earth's gravitational force on you is reduced. On the moon, your weight automatically would be about one-sixth of your weight on Earth, even though your mass remains the same.

Optics studies the movement of light. Progress in optics has helped to create more powerful telescopes, microscopes, fiber optics, lasers, and many other brilliant discoveries.

Classical thermodynamics studies how pressure and temperature affect the movement of matter and energy. Thermodynamics consists of four thermodynamic laws. These discoveries are expressed in mathematical formulas called equations of state.

The zeroth law of thermodynamics tells us that when we bring two systems together and leave them alone, they will eventually reach an equilibrium and have the same temperature. This implies that temperature is a universal physical quantity to describe the state of everyone and everything. Because of the zeroth law, we know we can use a thermometer to measure temperature and compare temperatures.

The first law of thermodynamics introduces the concept of energy into physics. It expresses the law of conservation of energy. It tells us that everything has an internal energy. You can change the internal energy by doing work, such as lifting a weight, or exchanging heat. You may also use the internal energy to do work or give heat. Energy can transform into different forms. It can also be transferred to another location. However, the total energy remains the same. The law of conservation of energy tells us that energy cannot be created or destroyed.

The second law of thermodynamics tells us that heat cannot spontaneously flow from a colder place to a hotter one. It introduces entropy as a physical quantity to describe the state of everyone and everything. Entropy measures the disorder within a system. Usually, the hotter the system, the more disorder it has. Therefore, a system with higher temperature has greater entropy than one with lower temperature.

Heat is the energy related to entropy. The second law of thermodynamics tells us that if left alone, a system will always become more disordered. Its entropy will increase and eventually reach a maximum. The second law of thermodynamics presents a phenomenon of irreversibility in nature. For example, if you don't put any effort into putting order into your room, it will become more and more disordered. It will not get neater by itself. Its entropy will increase.

The third law of thermodynamics tells us that as we lower the temperature of a system, its entropy gets smaller and reaches a minimum as the temperature approaches absolute zero.

Electromagnetism studies the movement of everything under the electromagnetic force. Developed over the course of the nineteenth century, classical electromagnetism culminated in the work of James Clerk Maxwell. Maxwell unified preceding developments in electromagnetism via a set of equations now known as Maxwell's equations. Through these equations, people discovered that light is an electromagnetic wave. The development of electromagnetism has led to the invention and use of electric lights, electric tools and machines, the telegraph, and many other applications.

For physicists used to Newtonian mechanics, electromagnetism appeared to be peculiar. In fact, it is not consistent with classical mechanics. According to Maxwell's equations, the speed of light in a vacuum is a universal constant. This violates the assumption in Newtonian mechanics that all laws of motion are identical in all inertial frames of reference under Galilean transformation. An inertial frame of reference is a frame of reference that describes time and space homogeneously, and in a time-independent and space-independent manner. A Galilean transformation treats space and time as separate identities. The assumption that space and time are independent of each other was a long-standing cornerstone of classical mechanics.

Modern Physics

The early twentieth century ushered in many new beginnings and dramatic changes for physics. New discoveries in physics started to challenge classical physics. New ideas and concepts have been introduced. Modern physics, which includes relativity theory and quantum physics, started to take shape.

Relativity theory

To reconcile electromagnetism and classical mechanics, Albert Einstein introduced special relativity. A defining feature of special relativity is the replacement of the Galilean transformations of Newtonian mechanics with the Lorentz transformation. In the Lorentz transformation, time and space are connected. A change in time will lead to a change in space, and vice versa. Space and time are interwoven into a single continuum known as spacetime. The interconnection between space and time yields the famous equivalence formula of mass and energy, $E = mc^2$.

Special relativity theory makes classical mechanics compatible with classical electromagnetism. Classical mechanics is the special case of special relativity theory when the speed of the moving object is much smaller than the speed of light.

Special relativity further shows that in moving frames of reference, a magnetic field transforms to a field with a nonzero electrical component and vice versa. This revealed that the electrical and magnetic forces are just two different aspects of one force, electromagnetism.

This great discovery of the oneness of the electrical force and magnetic force inspired Einstein. He spent much of his later years to continue this process of "unifying" forces by trying to unify the electromagnetic force with gravity. He did not live to see this effort come to fruition. To this day, unifying the electromagnetic force with gravity is an important ongoing pursuit in physics.

Special relativity theory is limited to flat spacetime. To address curved spacetime, wherein spacetime is curved and bent by mass and energy, Einstein created general relativity theory. General relativity is the theory about the relationships among matter, gravity, and spacetime. It describes how matter determines spacetime and how spacetime directs the movement of matter.

Einstein's theory has important astrophysical implications. For example, it implies the existence of black holes—regions of space in which space and time are distorted in such a way that nothing, not even light, can escape. It demonstrates that a black hole is an end-state for massive stars. General relativity is also the basis of current models of the expanding universe.

As if Einstein's relativity theory did not shake up classical physics enough, quantum physics dramatically challenges classical physics and the foundation of all natural science at its deepest core concepts and principles.

Quantum physics

Quantum physics studies what everything and everyone are made of and how they behave at the microscopic level. When physicists explore the world at smaller and smaller spacetime scales, or at higher and higher energy levels, they enter the quantum world.

Studies in quantum physics indicate that everything is made of various vibrations, also called waves. A vibration or wave is a periodic oscillation. Because quantum vibrations are not limited by space and time, everything is basically a vibrational field consisting of different vibrations.

A vibrational field is described mathematically by a wave function. A wave function is a mathematical formula that expresses the types and quantities of vibrations or waves within a system.

Quantum physics challenges the foundations of physics and natural science at their deepest core in three ways.

First, quantum physics is fundamentally non-deterministic. In quantum physics, everything is described by wave functions, which can only tell us the chances that certain things will happen. Because of the probabilistic nature of quantum physics, some scientists, including Albert Einstein, have been reluctant to accept quantum physics as the fundamental physics theory.

The measurement problem in quantum physics involves the debate about why our world appears to be definite while its underlying quantum nature is the superposition of many possible vibrational states. The main issue posed by the measurement problem is how observed reality is manifested from the vibrational field, which contains many possible states described by the wave function.

Second, the concept of quantum phenomena drastically conflicts with one of the cornerstones of natural science and scientific research, which is objectivity. It is generally accepted in natural science that natural phenomena are objective. Their existence does not depend on the actions of the observer. In quantum physics, however, phenomena are subjective and depend on the actions of the observer.

Third, some quantum phenomena, such as quantum entanglement or quantum correlation, challenge our concept of space and time. Two or more quantum waves can be quantum entangled when they are created from the same source. For quantum-entangled vibrations, if you do something to one of them, the other will be affected instantly and transform to a state that is determined by the quantum entanglement, no matter how far apart they are. This non-local effect is beyond the normal conception and perception of space and time. Albert Einstein called this quantum phenomenon a "mysterious action at a distance." This non-local effect contradicts the limitation on the transmission of information in Einstein's theory of relativity.

These three distinguishing qualities of quantum physics have caused many scientists to question whether quantum physics is a complete or correct theory about reality. Many interpretations have been proposed to make sense of quantum physics, such as the Copenhagen interpretation, pilot-wave theory, collapse theories, many-worlds interpretation, and more. Interpretations of quantum mechanics are attempts to comprehend quantum physics in terms of our ordinary understanding, as well as in philosophical and metaphysical terms.

The generally accepted Copenhagen interpretation of quantum physics discards the idea of the objectivity of physical reality. It suggests that physics is a subjective discipline that deals with only our knowledge of physical reality.

Pilot-wave theory is also called de Broglie-Bohm theory or the causal interpretation. It maintains the deterministic and objective nature of reality by making up a configuration in addition to the wave function that exists even when unobserved.

The many-worlds interpretation asserts the objective reality of the universal wave function by suggesting that the wave function describes actual existing parallel universes, the multiverse.

Wolfgang Pauli, John von Neumann, and Eugene Wigner suggested that the subjective nature of quantum reality is due to the fact that quantum theory was about mind-matter interaction.

Despite all the controversies and confusions regarding the interpretation of quantum physics, quantum physics is the most fundamental physics theory devised so far. Quantum physics provides the most powerful mathematical tool that makes the finest predictions about natural phenomena. Although seemingly very different, classical physics is the special case of quantum physics at low energy or at the macroscopic level. Quantum physics makes the most accurate predictions about nature. It has expanded our knowledge of nature tremendously in areas such as chemistry, material science, atomic physics, nuclear physics, and particle physics, as well as astrophysics, cosmology, and more. It has led to significant inventions, including superconducting magnets, light-emitting diodes, lasers, transistors, semiconductors such as microprocessors, medical and research imaging such as magnetic resonance imaging and electron microscopy, and explanations for many biological and physical phenomena. It opened the door for many important new discoveries in science, such as the structure of DNA, new fundamental particles, new forces, dark matter, dark energy, and tremendous amounts of new information and progress in astrophysics and cosmology.

Theory of Everything

When quantum physics goes to the subatomic level, it becomes particle physics. Particle physics is a branch of quantum physics that studies the fundamental building blocks of nature. So far, it has discovered that our world is made of twenty-four elementary particles, the Higgs boson, and four fundamental forces (strong, weak, electromagnetic, gravity).

Except for gravity, these basic elements are classified and explained by the Standard Model. In the Standard Model, the strong, weak, and electromagnetic interactions are mediated by particles called gauge bosons. The different species of gauge bosons are gluons, weak bosons, and photons. The Higgs boson appears in the Standard Model to give mass to other particles. Because of its unique function, the Higgs boson is called the "God particle" by some people.

Although the Standard Model represents most of the experimental data well, it does not explain where matter and forces come from and how they are created. Furthermore, it cannot include gravity in a mathematically consistent way. A more powerful theory is needed.

Quantum physics has sparked rapid progress in astrophysics. Astrophysics is the branch of physics that studies the composition and nature of the heavenly bodies, such as the sun, extrasolar planets, stars, the interstellar medium, galaxies, and the cosmic microwave background. Our knowledge about these heavenly bodies is increasing quickly with the development of more and more detectors.

"The more you know, the more you know how little you know." This applies perfectly in the field of astrophysics. Astrophysicists have found that matter as we know it makes up less than five percent of the universe. More than sixty-seven percent is dark energy, and about twenty-seven percent is dark matter. Dark energy and dark matter are basically energy and matter that we do not understand and cannot describe, other than that they should exist according to experimental data.

Current astrophysicists are striving to determine the properties of dark energy, dark matter, and black holes, the possibility of time travel, and other equally exciting topics about outer space. Rapid growth in astrophysics has propelled physical cosmology to a new level.

Cosmology is the branch of physics that studies the origin, evolution, large-scale structures, dynamics, and ultimate destiny of the universe, as well as the scientific laws that govern the universe. Einstein's general relativity provides an excellent mathematical tool to describe and study the large structures of our universe. Quantum physics provides the tools to delve deeply into what the planets, stars, galaxies, and universes consist of. Information about our universe is growing exponentially. Much observational data about our universe indicates that our universe started from a big bang, followed almost instantaneously by cosmic inflation, an expansion of space from which the universe is thought to have emerged 13.799 billion years ago. However, cosmology has not been able to tell us the source of our universe, nor how it is created, evolves, and ends.

Quantum physics provides the mathematical framework to unify the strong, weak, and electromagnetic forces, but failed when it tried to include gravity. Although quantum mechanics is not inconsistent with general relativity, they are definitely not compatible with each other. The search for the theory of everything (TOE) has been a major goal of twentieth and twenty-first century physics. Many prominent physicists, including Albert Einstein and Stephen Hawking, have labored for many years to find TOE. Stephen Hawking personally concluded that TOE is not obtainable.

String theory is one of the most promising candidates for TOE. String theory studies the quantum dynamics of a string. It is found that vibrations of a string may produce all the particles and forces observed in nature, including gravity. In string theory, all forces and all matter are naturally unified. However, current string theory is unable to make many testable predictions. Something is still missing in string theory. Many questions remain to be answered before string theory can become a theory of everything.

Physics is one of the highest achievements of humanity. As impressive and outstanding as physics is, current physics has two shortcomings. One is that it only deals with the material realm. It does not include spiritual existence. The other is that it does not answer the questions of where our universe comes from, how it is created, and how it ends. Greater unification of all theories is needed. Deeper understanding about creation is needed. Tao Science is born.

Tao Science

Tao is an ancient Chinese word that has many meanings, such as the way, road, course, principle, method, and more. In this book, we have used Tao to mean the Source and Creator, as Lao Zi did in his classic treatise, *Dao De Jing.* Tao Science is the science about creation. It integrates profound Tao wisdom with quantum physics.

Tao Science includes three laws:

1. Law of Shen Qi Jing
2. Law of Karma
3. Law of Tao Yin Yang Creation

The Law of Shen Qi Jing addresses the fundamental question of what everyone and everything are made of. This law teaches us that everyone and everything are made of shen, qi, and jing. Jing is *matter*, which is the physical reality we observe. Qi is *energy*, which is our ability to do work. Shen includes *soul*, *heart*, and *mind*. Here, soul is spirit. Heart is the *spiritual heart*. Mind is *consciousness*. Tao Science gives soul, heart, and mind (collectively, shen) scientific and mathematical definitions. Shen is *information*. Information has three aspects: content of information, receiver of information, and processor of information, which correspond to soul, heart, and mind, respectively. This definition makes it possible to study soul, heart, mind, as well as spiritual and conscious phenomena, scientifically and mathematically with quantum physics. In this way, Tao Science brings natural science and spirituality together at the most fundamental level. Understanding the Law of Shen Qi Jing empowers us to gain a deeper and better understanding of ourselves, everyone, and everything, as well as the deeper meaning and greater purpose of our lives. It also teaches us what our higher powers are and how to develop and use our higher powers to transform and uplift every aspect of our lives.

The Law of Shen Qi Jing also teaches us the profound wisdom and practice about what the highest spiritual state is: the ultimate enlightenment, immortality, bliss, and freedom—and, most important, how to reach it.

The Law of Karma reveals scientifically how our current experience is affected by our past actions and how our current actions create our future reality. It shows us the root cause of everything. The Law of Karma empowers us to heal and transform every aspect of our lives and become a more powerful creator and manifestor.

The Law of Tao Yin Yang Creation is the fundamental law about creation. It reveals how everyone and everything and our entire universe are created. From the Law of Tao Yin Yang Creation, we can derive string theory, M-theory, and all the current physics theories, as well as ancient Chinese wisdom, including the *I Ching* and Five Elements (Wu Xing) Theory. The

Law of Tao Yin Yang Creation holds the key to help us reach ultimate enlightenment and freedom.

Tao Science is the science about creation. It is to empower us to create the life we truly want. Tao Science is the science of grand unification. It brings grand unification to everyone and everything. Tao Science is the science to attain our higher life potentials and purpose. Tao Science is the science to gain greater abilities and powers. Tao Science is the science to reach enlightenment, immortality, and ultimate bliss and freedom.

3

Law of Shen Qi Jing

IN THIS CHAPTER, we introduce the first foundational law of Tao Science, the Law of Shen Qi Jing. The Law of Shen Qi Jing is based on profound ancient Chinese wisdom about what everyone and everything are made of. We will use modern scientific terms to express and explain the ancient wisdom. As you will learn, the Law of Shen Qi Jing holds the key to unveil the secrets about who we really are, the true purpose and meaning of our lives, our higher powers and abilities, and how to develop and use those abilities. It unlocks a gate to bridge and integrate natural science and spirituality. It also brings a new approach to explaining quantum physics to the lay person in simple, accessible, and easily comprehended terms.

What Are We Made Of?

Have you ever wondered who we are? What are our higher abilities and potentials? What is the deeper meaning and purpose of our lives?

Many of us have asked these questions. Many of us have searched for the answers to these questions. To gain deeper insight into these important questions, we need to explore what everyone and everything are made of.

Throughout history, there have been many different ideas about what everyone and everything are made of. Natural science tells us we are made of matter and energy. It studies the materialistic nature of our existence. Social science, philosophy, psychology, and spiritual and religious ideology state that our essence is spirit and consciousness. They account for the spiritual and conscious aspects of our existence.

According to ancient Tao wisdom, everyone and everything is made of jing, qi, and shen. Jing is matter. Qi is energy. Shen includes soul, heart, and mind. Soul is the spirit. Heart includes both the physical heart and the spiritual heart. Mind is consciousness.

This simple and yet profound Tao wisdom is the first law in Tao Science, the Law of Shen Qi Jing.

Law of Shen Qi Jing

Everyone and everything are made of shen (soul, heart, and mind), qi (energy), and jing (matter).

The Law of Shen Qi Jing provides the breakthrough insight needed to solve many difficult problems in science. For example, many of us know that scientists have been trying to reach a completely satisfactory understanding of the fundamental nature of reality as presented by quantum physics. This law yields a new, simple, clear, and profound way to understand and interpret quantum physics in terms everyone can grasp. It paves the way to integrate science and spirituality naturally into one at the most fundamental level. It also lays the foundation to derive the Grand Unification Theory.

Even more exciting, this law is highly significant for you, me, all humanity, and all beings. It reveals to us our highest self, the much greater potential and power within each one of us, and the deeper meaning and higher purpose of our lives.

Now let's explore how all of these come about.

What Are Soul, Heart, and Mind?

To understand the Law of Shen Qi Jing scientifically, we need to give shen a scientific definition. First, what are the components of shen—soul, heart, and mind? Different people, cultures, traditions, ideologies, religions, philosophies, and sciences have different understandings.

While attending one of Master Sha's workshops, Dr. Rulin had an "aha!" moment. She saw that it is possible to define soul, heart, and mind as physics quantities and calculate them in quantum physics. Consequently, we can understand and study soul, heart, and mind, as well as spiritual and conscious phenomena, scientifically and mathematically. This insight eventually led to Tao Science.

In Tao Science, soul, heart, mind, energy, and matter are defined as follows:

Soul is the content of the information in everyone and everything.

Heart includes the physical heart and the spiritual heart. The spiritual heart is the receiver of the information in everyone and everything.

Every system, every organ, every cell, every DNA, every RNA, and every smallest matter has its spiritual heart. In this book, when we speak of "heart," we are generally referring to the spiritual heart.

Mind is the processor of the information in everyone and everything.

Energy is the ability to do work, such as lifting a weight. Energy is the actioner of everyone and everything.

Matter is the physical reality. It is everything we can observe and measure: weight, length, height, charge, mass, electrical field, shape, color, frequency, and other physical properties and quantities. **Matter is the transformer of everyone and everything.**

In Tao Science, the manifestation process is:

Soul → Heart → Mind → Energy → Matter

Soul gives a message to the heart. Heart receives the message and activates the mind. Mind processes the message and directs the energy. Energy takes action and moves the matter. Matter is what we experience. Our soul, heart, mind, and energy determine the matter, which is our physical reality. On the other hand, a change in matter can transform our soul, heart, mind, and energy. Therefore, matter is the transformer. Our physical life is to serve our spiritual journey. It is to transform our souls, hearts, and minds.

Now let's explore in further detail the above definitions and processes.

What Is Information?

We are in the Information Age. We are dealing with information in every moment and every aspect of our lives. Information affects our lives in a profound way. We all know the importance of information. We are familiar with the fact that the information in financial institutions and government records determines our wealth. For instance, a family owns a large piece of land with a big mansion for generations. This family is considered to be wealthy. If one day the information changes and indicates that the land and mansion are not worth much and this family owes a huge amount of taxes and carries other debts, the family is then financially poor. This shows us that information determines our wealth. Good information is of the highest value.

The importance of information cannot be overestimated. For example, this fact is very clear during a war. One piece of information can save or destroy an entire country. Information is essential for every aspect of our lives. Information can light up or ruin a relationship. It can bring us joy; it can also bring us worry or grief. It can make us sick or healthy. It can enlighten us; it can confuse us. It can bring fortune as well as disaster to our lives.

What is information? Information is the message. It is that which informs. To inform is to answer a question. Questions can be posed so that the answer is either "yes" or "no." Therefore, information can be represented as a sequence of "yes" and "no." A computer specifies information by a series of 0s and 1s.

The mathematical definition of information is a relatively recent development. In the 1920s, scientists at Bell Labs studying how to improve the transmission of information by telegraph found it necessary to define information as a measurable mathematical quantity. Claude Shannon, the founder of Information Theory, realized that to measure information is to count the number of possibilities. He used bits to measure information. One bit of information refers to two possibilities—yes or no, black or white, and so on. Two bits of information refer to four possibilities, two pairs of black or white, or two pairs of yes or no. Three bits of information refer to eight possibilities, three pairs of yes or no. For instance, a coin has two sides. If you put one coin on a table, there are two possible images. In mathematics, we say there are two possibilities (2^1). You need one piece of information to determine the state of the coin. If you place two coins on a table, there are four possible states, four possibilities (2^2). You need two pieces of information to determine the state of the two coins.

In the general case, Shannon found that the measure of information relates to the entropy discovered in thermodynamics. Entropy measures the possible states in a system. Specifically, it measures disorder. Disorder is the existence of unrelated possible states. A fundamental discovery in thermodynamics is that heat, which is one form of energy, increases with entropy and temperature. This tells us that information can directly create energy. Heat is the energy created by information.

Matter and energy can be processed, transported, manipulated, and transmitted; so can information. Living in the Information Age, most of us are familiar with processing, transmitting, downloading, moving, and transferring information. Computers and the internet are technologies that make it easy for us to process, transfer, transmit, move, and download information. It has become an important part of our lives.

Our world, everyone, and everything are made of matter, energy, and information. Matter, energy, and information coexist in everyone and everything. There is no matter that does not carry information and energy. Similarly, there is no energy that does not carry matter and information. And there is no information that is not carried by matter and energy. Matter and energy are the carriers of message or information.

Matter, energy, and information are the three basic elements of everyone and everything.

We suggest that shen is information. Qi is energy. Jing is matter. The Law of Shen Qi Jing can also be expressed as:

Everyone and everything are made of information (shen), energy (qi), and matter (jing).

Soul, heart, and mind are three aspects of information: the content of information, the receiver of information, and the processor of information.

Quantum Physics of Shen Qi Jing

Quantum physics is the science that currently reveals at the deepest level what everyone and everything are made of and how they behave. What can quantum physics tell us about our soul, heart, mind, energy, and matter? The quantum world is intriguing, magical, and empowering. We will also see how the Law of Shen Qi Jing can shed light on some challenging problems in quantum physics.

Imagine now you are a quantum physicist. You have a powerful microscope. You are going to use this microscope to find out what everyone and everything are made of and how they work at the smallest scale.

You look at your body under the microscope. You find that although different parts of your body look different, they are all essentially the same in the sense that they are all made of cells.

Now you adjust your microscope and look at the cells. You find that although there are many kinds of cells, all the cells are made of molecules.

Now you adjust your microscope and look at the molecules. You find that although there are many kinds of molecules, all the molecules are made of atoms.

Now you adjust your microscope and look at the atoms. You find that although there are many kinds of atoms, all atoms are made of a nucleus and electrons.

Now, because your microscope is so powerful, you can adjust your microscope and look at the nucleus. You find that the nucleus is made of protons and neutrons. In real life, no existing microscope is powerful enough to see what is going on inside a nucleus. Particle physicists study the nucleus by using an accelerator to produce particles with high energy. They "bomb" the nucleus with these high-energy particles, and then observe what is produced by the "bombing." From this analysis, they have found that the nucleus is made of protons and neutrons. In the same way, they study what protons and neutrons are made of. Through long and difficult experiments and mathematical deduction, they have found that protons and neutrons are made of quarks and gluons.

Now you look at quarks, electrons, gluons, and photons. You find that they seem to be complete. You can't break them down further. They are the basic building blocks for our world. Quantum physicists call them elementary particles.

To your amazement, although these elementary particles have energy, charge, spin, and even mass, they don't behave like particles at all. They all behave like waves.

Wave is also called vibration. Wave is a periodic oscillation. Wave is characterized by its wavelength, frequency, and amplitude. Wavelength is the distance between successive crests of a wave. It is the distance between two adjacent waves. Frequency is the number of occurrences of a repeating oscillation per second. Frequency measures how quickly a wave oscillates. Period measures the time it takes a wave to oscillate for one complete cycle. Amplitude measures the height of a wave. A wave is constantly moving. When a wave moves, it carries matter, energy, and information to different places. Velocity measures how fast a wave travels.

The following diagram illustrates the wavelength, frequency, amplitude, and velocity of a wave:

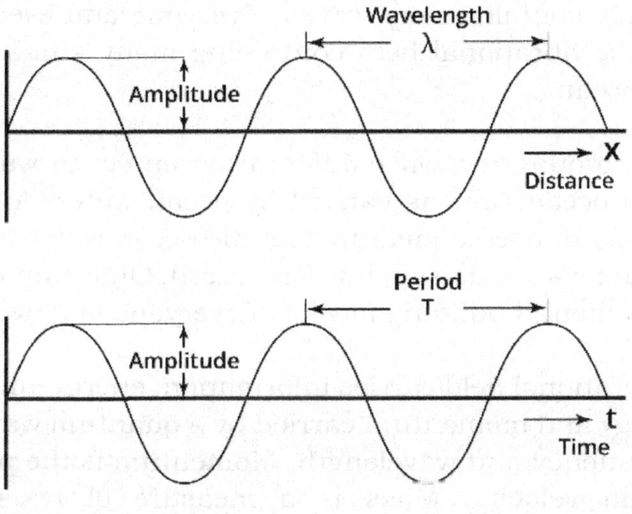

Figure 1. Characteristics of a wave

Now let's sit at a beautiful beach in Hawaii and watch the ocean waves. You can jump into the ocean and let the waves carry you up and down. You can surf the waves and let the waves move you forward. You can simply watch the waves and surfers, seeing the excitement and exhilaration of surfers catching and riding the waves.

Observe the ocean waves closely. Unlike a stationary object, waves never stay still in one location. They constantly oscillate and travel. The ocean waves oscillate, going up and down. They travel, splashing at the shore. They move from one place to another. They exist in the whole ocean. In the same way, a quantum wave constantly oscillates and moves like the ocean waves. Quantum waves exist in all space and time.

Now look at the ocean waves more closely. You find there are all kinds of waves in the ocean. There are long waves and short waves. There are tall waves and low waves. Like the ocean, everyone and everything does not usually contain only one wave. They contain many waves. Everyone and everything is basically a vibrational field containing many kinds of waves like the ocean.

However, a quantum wave differs from an ocean wave in two ways. An ocean wave is carried by ocean water. A quantum wave does not need a medium like the ocean water to carry it. A quantum wave exists on its own accord. Quantum waves are the fundamental building blocks of everyone and everything.

Every vibrational field carries information, energy, and matter. The energy and momentum carried by a quantum wave relates to its frequency and wavelength. Momentum is the product of mass and velocity. Mass is a measure of resistance to movement, or inertia. Velocity is the quantity that tells us how fast something moves. The more momentum someone or something has, the bigger impact you will feel when you bump into them. Quantum physicists find that the higher the frequency a wave has, the more energy it contains. The shorter its wavelength is, the more momentum it carries.

Wave Function

Since everyone and everything is made of waves, quantum physicists use wave functions to mathematically describe everyone and everything. A wave function is the mathematical formula that accounts for the kinds of waves and the amount of each wave one has in one's vibrational field.

The most common symbols for a wave function are the Greek letters ψ or Ψ (lower-case and capital psi). When expressed in terms of space-time coordinates, the wave function tells us the probability that the object will be at the appointed space and time. From the wave function, we can calculate the properties and qualities, such as information, energy, frequencies, and matter, about everyone and everything. From the wave function, we can calculate our soul, heart, and mind.

In summary, quantum physics tells us that at the deepest level:

Everyone and everything is a vibrational field. The vibrational field carries information (shen, including soul, heart, and mind), energy (qi), and matter (jing).

A vibrational field is something that extends over space and time. When we look deeper and deeper into everyone and everything, we discover that:

- We are not as solid as we appear to be. Our vibrations can travel through walls.
- We are not the limited physical objects we appear to be. We are infinite fields extending over all space and through all time.

Ancient Chinese writings reveal that Taoist saints had the ability to disappear, travel with the wind, appear at a faraway place instantly, travel through Heaven, walk through walls, and much more. Tao Science can explain the scientific possibility of these abilities because everyone and everything is essentially a vibrational field. The natural question is: why could these Taoist saints possess such abilities while ordinary people do not? Can we develop these abilities? How can we do that? One goal of Tao Science is to empower you with the wisdom and practices to achieve these higher powers and abilities. We will continue to explore these questions. As you will learn, the answer to these questions lies in the power of positive shen qi jing.

Everyone and Everything Has Soul, Heart, and Mind

According to our definitions above of soul, heart, and mind, everyone and everything has soul, heart, mind, energy, and matter. Quarks, electrons, atoms, molecules, cells, animals, plants, mountains, water, minerals, planets, stars, galaxies, and universes all have soul, heart, mind, energy, and matter. Nothing exists without soul, heart, mind, energy, and matter. This is because everyone and everything carries information, energy, and matter. Everyone and everything is constantly receiving and processing information, energy, and matter. In this sense, everyone and everything has consciousness.

Researcher Cleve Backster spent thirty-six years studying bio-communication in plant, animal, and human cells. Formerly an interrogation specialist for the CIA, Backster utilized polygraph equipment to monitor plant, animal, and human cells in his laboratory. He found that plants became attuned to their primary caretakers and responded to both their positive and negative emotions. Backster noted that the distance between the plants and their caretakers seemed to be irrelevant in these experiments. This indicates that the connection between plants and their caretakers transcends space and time. Even blocking electromagnetic radiation from and to the plants did not break this connection. This means that this connection is not due to transmission of information through the electromagnetic field. In similar experiments, Backster discovered that the same type of connection exists between white blood cells and their human donor. The spontaneous emotions of the donor can affect the cells' activities regardless of the distance between them. Backster's experiments demonstrate that plant, animal, and human cells have consciousness.

Many traditions and cultures have the wisdom and knowledge that everyone and everything has spirit, heart, and consciousness. This wisdom and knowledge is reflected in every aspect of Hawaiians' lives. In Hawaii, you will hear a story about how a small pebble can ruin your relationship and bring disasters to your life if you do not treat it right. People consult with the land before they do any work on it. For example, they ask a tree for permission before they cut it down. When lava flowing from a volcanic eruption forces an evacuation, they put whisky and a well-prepared meal in front of their house to show their honor and respect to Pele, the Fire Goddess of the volcano. There are many stories about how this gesture has saved many houses from destruction.

Living in Hawaii, Dr. Rulin learned from Hawaiians how to live in this magical world where all beings are conscious. She can never forget the first time she heard a tree talking to her. She will always remember the bliss she felt when dolphins came to her to share their joy, wisdom, and sacredness. She will forever treasure the immensely blissful experience when a boulder near the ocean next to her home shared its life journey of millions of years with her. She will be eternally grateful for the teachings, blessings, and downloads she has received from Pele and other Hawaiian gods and goddess, as well as from sacred sites such as Mt. Shasta, Sedona, Mexican pyramids, and from many spiritual fathers and mothers, including Master Sha, Babaji, buddhas, Taoist saints, and other holy beings.

In her own words, she shares some of her experiences with Madam Pele, the Hawaiian goddess of the volcano:

“Two Hawaiian kahunas took us to a volcanic crater. Kahunas are Hawaiian priests. They carry the sacred wisdom, ability, and practice to communicate and interact with the spiritual world. They held a simple ceremony, including an invocation and Hawaiian chant, to connect with Pele. Then they guided us to go on a spiritual journey to meet with her.

“Pele appeared to me in spiritual form as a beautiful woman with lavish dark black hair and a body filled with hot vitality. She took me to her magnificent palace and showed me her immense creation power. She demonstrated how she could move Earth with her passion. Then she turned to me, looking deep into my eyes, and said, ‘My beloved sister, you have the same power. Trust yourself.’

“After this experience with Madam Pele, every full moon night, I will take a hike in the moonlight without a flashlight to see her. Each time, she shares a lot of wisdom and gives me a lot of blessings.

“Pele loves me dearly as her beloved sister. She always treats me with amazing hospitality. Sometimes, she tries to scare me and test me. Once, a spiritual master invited me to do a ceremony with him to take people’s offerings to Pele. We walked on a lava field, the hot orange and red lava right under our feet with only a thin layer of black lava rock supporting us. Wearing simple sandals with most of my feet exposed, the idea came to my mind that I could easily lose both of my feet if only a small portion of the red and orange lava touched them. Trusting Pele’s love, I walked on. We did a beautiful offering ceremony to Pele.

"Once I took one of my guests to see Pele. We took a long walk on the black lava field. We finally reached the ocean where some hot red orange lava poured into the ocean just two feet below us. We did a simple ceremony expressing our gratitude to Pele and Mother Earth. I offered one of the two apples we brought to Pele. Then, we sat on a stone quietly eating our supper.

"Suddenly, I noticed the lava flow under us was a lot larger than before. It continued to grow fast. Before we realized what was going on, a large river of lava appeared, flowing to the ocean. I started to yell, 'Pele is giving us a show! Pele is giving us a show! Pele is giving us a show!'

"As if encouraged by my enthusiasm, the lava river increased quickly and became a lava lake. Then, a lava 'waterfall' was added. My friend and I cried, shouted, and screamed with great excitement like little kids. Loving our fervor, Pele unleashed many more lava falls! 'Wow! Wow! Wow!' We had never seen anything as spectacular as this before. We were yelling with more and more gusto.

"Further stimulated by our zeal, Pele let out a big splash of red orange lava in the air. It looked like fireworks. The splendor was beyond words. We were amazed and enchanted.

"After more than an hour of the greatest excitement, the two of us were exhausted. We told Pele how impressed and grateful we were for the show she gave us. We sat there quietly, still enjoying the spectacular scene. Then, we saw the lava fireworks and lava falls disappear. The lava lake gradually got darker and darker, and then sank.

"We suddenly realized we should take a photograph. We brought out our camera. The camera's battery was discharged and we could not even take one picture. I realized that Pele does not want us to show off her power to others. She is my sister. She simply wanted to give me and my guest a treat for coming to visit her."

– Dr. Rulin Xiu's experience with Pele,
the Fire Goddess of the volcano, in 2008

When you open your soul, heart, and mind and connect with the souls, hearts, and minds of all beings, your life will be expanded beyond your comprehension. You will start to experience a world with immense joy, love, wisdom, harmony, abundance, and power. Beloved reader, we hope Tao Science will open your soul, heart, and mind and empower you to live a life full of love, joy, wonder, power, and miracles.

Unification of Science and Spirituality

With the full flowering of the scientific revolution some three hundred years ago, natural science was separated from the spiritual, conscious, and religious disciplines. Physics, as the foundation of natural science, studies matter and energy. It uses quantities such as mass, weight, volume, energy, velocity, entropy, electric field, spin, and more to describe everyone and everything. Quantities in physics are the items that can be measured physically and calculated mathematically. Physics laws use mathematical formulas to express repeatable experiments about matter and energy. Because of the repeatability and calculability of these quantities, physics and natural science have the great ability to create inventions to utilize, transform, and transport matter and energy. Physics and natural science have helped people live better physical lives. The resulting inventions and new technologies have made it much easier to perform physical work and freed us from many kinds of labor. They have made many dreams and fantasies a reality. Now, we can fly to almost anywhere on Mother Earth and even to outer space. We can speak almost instantly with anyone on Mother Earth who also has a phone. We can see deeply into space, billions of light years from us.

However, the focus of natural science on physical existence to the exclusion of our soul (spirit), heart, and mind (consciousness) has had serious negative side effects. It has led many people to become slaves of the material world. It has caused deep segregation within our own souls, hearts, minds, and bodies as well as in our cultures, societies, and world. With the ever-increasing popularity and impact of science on our lives, this separation has contributed to the alarming increase in depression and anxiety, especially among children, and to escalating challenges in all our relationships. It has produced unprecedented pollution in our souls, hearts, and minds, in our societies, in our environment, and in the world. It has helped cause large-scale wars and massive destruction to nature. Science without soul, heart, and mind has proven to be dangerous again and again. "Progress" in natural science has led to the creation of nuclear, chemical, and biological weapons, some of which can wipe out humanity in a matter of minutes.

On the other hand, spirituality and religion without science are also incomplete. As Albert Einstein said:

"Science without religion is lame, and religion without science is blind."

Some people have lost faith in spiritual beliefs and traditions.

The unification of science and spirituality is urgently needed to make us, our society, and the world whole again. It can help bring more love, peace, and harmony to humanity and Mother Earth. Our dream is that science and spirituality can join as one to enrich each other and uplift each other.

In recent years, more and more people have searched for the way to integrate science and spirituality. Many have attempted to explain spiritual and conscious phenomena scientifically in terms of matter and energy. Some have tried to account for physical phenomena with consciousness and spirituality. Many philosophies, ideologies, and traditions presume the coexistence of physical and spiritual existence.

The Law of Shen Qi Jing enables us to study soul, heart, mind, energy, and matter together within one scientific framework. It provides a way to integrate natural science with spirituality at the most fundamental level. As we will show later, the Law of Shen Qi Jing provides a way to give a simple yet powerful metaphysical interpretation of quantum physics. It also sets the foundation for the Grand Unification Theory to unify not only the fundamental physical forces, but also to unify our souls, hearts, minds, and bodies, to unify science with love and spirituality, and to unify humanity with nature, so that we can live harmoniously with ourselves, each other, and our environment. More importantly, this law makes it possible to transcend our human limitations, to connect and uplift each individual and all humanity to our higher powers.

In summary:

- The Law of Shen Qi Jing (in scientific terms, the Law of Information Energy Matter): everyone and everything is made of shen (information), energy, and matter.

- Shen is information. Information is the possibilities and possible states of an entity.

- Shen includes three elements: soul, heart, and mind. Soul, heart, and mind relate to three aspects of information.

  - Soul is the content of the information of an entity.
  - Heart is the receiver of the information of an entity.
  - Mind is the processor of the information of an entity.

- Everyone and everything has soul, heart, mind, energy, and matter.
  - Energy is the ability of an entity to do work.
  - Matter is everything we observe in the physical realm.

An important personal practice is to observe what kind of information is carried in your thoughts, feelings, emotions, speech, and more. Be aware of how you receive and process information. See how the information affects your life.

4

Power of Soul

SOUL IS THE CONTENT of information within one's vibrational field. One's soul determines one's heart, mind, energy, and matter. Soul is the boss. To know the power of the soul is to understand the highest purpose of our lives and to utilize the highest power we have.

Soul Light Era

The beginning of the twenty-first century is the transition period into a new era for humanity, Mother Earth, and all universes. This era is named the Soul Light Era. The Soul Light Era began on August 8, 2003. It will last at least fifteen thousand years.

The natural disasters, human catastrophes, social upheavals, wars, terrorism, diseases, nuclear weapons, pollution, economic challenges, energy crises, vanishing species, global warming, and many other phenomena we are now experiencing are part of this transition.

On the individual level, the number of people who suffer from pain, depression, anxiety, fear, anger, grief, and worry, as well as chronic diseases and relationship and financial challenges, is escalating at an alarming rate. All of these are signs that humanity needs to transform at a deep level to a new way of being.

Humanity has gone through many stages of history with different levels of consciousness. The first level of consciousness is **survival consciousness**. In this stage, people focus on physical survival. They use human labor, natural resources, animals and plants, marriage, human reproduction, wars, laws, and more to survive and propagate their beliefs and cultures. This period of human history is dominated by fighting for land, wealth, physical resources, and many other material objects. Let's call this the era of "force over matter."

The second level of consciousness is **energy consciousness**. In this stage, people strive to cultivate their energy. They developed practices such as yoga, kung fu, qi gong, machines, tools, weapons, vehicles, domesticating animals, mining for coal, drilling for oil, cultivating nuclear power, nuclear weapons, medicine, and many other technologies, as well as laws, rules, and even wars, to enhance their ability to use energy. Let's call this the era of "energy over matter."

The third level of consciousness is **mind consciousness**. In this stage, people realize the power of the mind. Mind has the capacity to process information and direct energy. People use computers and many other instruments to enhance their mind power. They also use psychoanalysis, mind control techniques, yoga, meditation, kung fu, qi gong, and many other ways to control and purify the mind. They use media, propaganda, marketing and advertising, books, videos, movies, music, and more to influence people's minds. Let's call this the era of "mind over matter."

The fourth level of consciousness is **heart consciousness**. In this stage, people realize the power of the heart. The function of our spiritual heart is to receive information. We invented the telescope, microscope, accelerators, particle detectors, microwave detectors, infrared and ultraviolet spectrometry, gamma ray detectors, MRI, and many other instruments, as well as satellites, space shuttles, and more, to expand our ability to receive information. However, most of humanity has not fully realized the power of our own hearts and the deep profound truth about "heart over matter": *what our heart receives is what we manifest.* We have not opened our own hearts enough to receive high-level wisdom, knowledge, and messages. Heart disease is currently the number one cause of death in the world.

The fifth level of consciousness is **soul consciousness**. Recall that soul is the information within us. At this stage, we recognize our souls have many miraculous abilities, such as intuition, direct knowing, telepathy, clairvoyance, distance healing, teleportation, and more. We can use these soul abilities to achieve great success in every aspect of our lives. We realize the importance of our souls. Heal the soul first, and then healing of heart, mind, body, and every aspect of our lives will follow. We learn to use the power of our souls to heal, transform, and uplift our souls, hearts, minds, energy, bodies, and every aspect of our lives. We begin to see the greater potentials, the deeper meanings, and the higher purposes of our lives. We connect with, cultivate, heal, transform, and enlighten our souls. This is the era of "soul over matter."

The Scientific Revolution consummated the era of "force over matter". The Industrial Revolution consummated "energy over matter." The Information Age started the era of "information over matter." Tao Science will take us from the Information Age to the Soul Light Era. The Soul Light Era is the era of "soul over matter." In the Soul Light Era, humanity will reach a new level of existence with miraculous abilities, fully realized potentials, powers, intelligence, purpose, and meaning for life.

Let's now step into the Soul Light Era and explore the power of soul and how to use it.

Soul Holds Vast Wisdom and Knowledge

The power of soul consists of two aspects. One is higher wisdom and broader knowledge. The other is miraculous abilities and powers.

Soul is the content of our information. Our souls contain vast information. Some of the information is accumulated through our past actions, behaviors, speech, and thoughts, some from our ancestors, and some from our connections with Mother Earth, Heaven, and more. Accessing our vast soul information can provide us with massive and high-level wisdom and knowledge.

Many people have heard about the Akashic Records. Akasha is the Sanskrit word for *air* or *aether*. In Hindi, akash means *sky* or *Heaven*. The Akashic records are records of all events, thoughts, words, emotions, and intents that have ever occurred. Some people believe the Akashic records are encoded in a non-physical plane of existence known as the etheric plane. There are anecdotal accounts but no scientific evidence for the existence of the Akashic Records.

In Tao Science, everyone and everything has a soul. Soul is the content of information in our vibrational field. The wave function accounts for it mathematically. Your Akashic record is essentially the information within your soul.

A quantum vibrational field extends over all space and time unless it is blocked. The information of your soul extends over all space and time. Your soul information can be received over all space and time. Therefore, when you can do high-level soul communication, you can obtain your own as well as other people's soul information even when you are not in direct contact with them.

In addition to the vast information and wisdom held within your soul, your soul also has miraculous powers. To comprehend your soul's miraculous abilities, you need to understand an important quantum phenomenon: quantum entanglement.

Quantum Entanglement

Quantum physicists discovered an earth-shaking quantum phenomenon: quantum entanglement. When two or more waves are created from the same source, they are quantum entangled. Quantum entanglement means that the states of two or more waves or vibrations are connected. When one or more waves are observed at certain states, the other waves with which it is quantum entangled will instantly manifest into the state that is determined by the quantum entanglement, no matter how far they are from each other. This seems to violate the principle of cause and effect.

Albert Einstein called quantum entanglement phenomena "mysterious action at a distance" because it is instant. He thought it did not depend on the exchange and transfer of information through space and time. This does not agree with the basic principle of Einstein's theory of relativity. According to Einstein's relativity theory, information cannot travel faster than the speed of light. Because quantum entanglement phenomena are instant, they are impossible according to Einstein's relativity theory.

Many scientists have tried to disprove quantum entanglement. However, more than one hundred years later, quantum entanglement has been verified by numerous experiments as a real quantum phenomenon.

Quantum entanglement is an important property of quantum waves. Since everyone and everything is made of quantum waves, everyone and everything has this property. Everyone and everything can affect instantly the ones with which it is quantum entangled, no matter how far apart they are. How such non-local phenomena can happen is one of the conundrums of modern physics.

In chapter eleven, we will show you how to understand and derive quantum entanglement from the fundamental laws and principles of Tao Science. For now, let's simply accept it as a basic quality of our vibrational field.

The Miraculous Power of Soul

Through quantum entanglement, your soul can have instant impact and influence on others. Through quantum entanglement, your soul can travel from one place to many other places instantly. Through quantum entanglement, your soul can know at a distance, affect instantly at a distance, intuit, heal remotely and instantly, communicate telepathically, produce psychokinesis, possess clairvoyance, and more.

Many of us would never have thought that we have these miraculous soul abilities. However, we all have had the experience that we simply know certain things without anyone telling us. You may sometimes notice how your thoughts and feelings affect other people even when you keep them secret. Conversely, you may also notice how other people's thoughts and feelings affect you even if they try to hide them from you.

Everyone and everything is made of quantum waves. Quantum entanglement is one of the basic qualities of quantum waves. Therefore, everyone and everything essentially have these miraculous spiritual abilities. They are not special privileges or gifts of a few chosen spiritual masters. Each one of us has the capability to create miracles. It is our birthright. But we have to cultivate our souls, hearts, minds, and bodies to increase our abilities of quantum entanglement and soul power. Everyone and everything have the soul power to affect the ones with which they quantum entangled instantly. The more quantum entanglement we have with others, the more soul power we possess.

Russell Targ is a physicist who spent several decades working in a United States government program exploring "remote viewing." Remote viewing is the practice of seeking impressions about a distant or unseen target using subjective means, in particular, extrasensory perception (ESP). As cofounder of the Stanford Research Institute (SRI), a research and development think tank in Menlo Park, California, Targ investigated psychic abilities in the 1970s and 1980s. His research on remote viewing has been supported by the U. S. Central Intelligence Agency for twenty-three years, and has been published in *Nature,* the *Proceedings of the Institute of Electronic and Electrical Engineers* (IEEE), and the *Proceedings of the American Association for the Advancement of Science* (AAAS). His latest book (2012) is *The Reality of ESP: A Physicist's Proof of Psychic Abilities.*

In one of his experiments, Targ demonstrates that the evidencc of human's remote viewing ability is ten times stronger than the evidence of the efficacy of aspirin. Specifically, remote viewing has an effect size ten times greater than aspirin's. Effect size measures the relative power of experiments, and not just their probability. One of his experiments was to apply remote viewing to predict stock market performance. The subjects were able to make $120,000. Targ not only showed that people have remote viewing abilities, he also found that this ability can be improved with training.

For the past twenty years, Targ has been traveling and teaching people to do remote viewing. He has trained members of the U. S. Army to discover and improve their remote viewing abilities. He even created a free smartphone app, ESP Trainer, to help people enhance their remote viewing abilities. In other experiments, Targ and other scientists also found that strong telepathy exists between people who have a strong emotional bond. Targ concluded through his thirty years of research that remote viewing and other psychic abilities are natural abilities everyone possesses.

Soul power is miraculous. It is the power of the twenty-first century. Soul power is the new cutting-edge technology available to everyone and everything. It is critical that we cultivate and utilize positive soul power to help us achieve healing, prevention of sickness, rejuvenation, prolonging life, transformation of life and consciousness, and bringing love, peace, and harmony to the world in an unprecedented way.

Since 1998, I (Master Sha) have been devoted to training people worldwide to use their soul power to heal and transform every aspect of their lives. In my Soul Power Series of ten books, including the authority book, *The Power of Soul*,[4] and all of my books, I teach how to apply the power of the soul. The techniques are simple, practical, and profound.

What is the power of soul? Everyone and everything has the following qualities and characteristics of soul power:

- Your soul has great wisdom and knowledge. After you open your spiritual communication channels, you will be able to consult with your soul. You will be amazed to learn how much your soul knows. Your soul is one of your best teachers, consultants, and guides.

---

[4] New York/Toronto: Atria Books/Heaven's Library Publication Corp., 2009.

- Your soul has great memory. A soul can remember experiences from all of its lifetimes. For example, you may travel somewhere for the first time but clearly feel you are familiar with that place. You may feel like you were there before. Some places make you happy. Some places make you scared. You may have had past-life experiences in those places. Your soul has memories of those experiences. Therefore, you have special feelings at those particular places.

- Your soul has flexibility. Your soul holds infinite possibilities. There is never only one possibility or one choice; there are always many possibilities for everyone and everything in every situation. You always have the chance to choose the best possibility for you and everyone.

- Your soul communicates with other souls naturally and constantly. People often talk or dream about a soul mate. When you meet some people, you may instantly feel love. You may feel there is something special between you. The reason for this is that your souls were close in past lives. Your souls could have been communicating for many years before you met physically.

- Your soul travels. When you are awake during the day, your soul remains inside your body. But when you are asleep at night, your soul may travel outside your body naturally. In fact, many souls do this. Where does the soul go? It goes where it loves to go. Your soul can visit your spiritual teachers to learn directly from them. It can also visit your old friends, or Heaven and other parts of the universe.

- Your soul has incredible power for healing, including self-healing, healing of others, group healing, and remote healing.
- Your soul can help you prevent sickness and other challenges in your life.
- Your soul can help you rejuvenate.
- Your soul has incredible blessing capabilities. If you encounter difficulties and blockages in your life, simply ask your soul to help you: *Dear my soul, I love you, honor you, and appreciate you. Could you bless my life? Could you help me overcome my problems and difficulties? Thank you so much.* Invoke your own "body soul" in this way anytime, anywhere. Your soul can help you solve your problems and overcome your difficulties. Love your soul. Ask your soul to bless your life. Your soul will be delighted to assist you. You could be fascinated and amazed to see the changes in your life.
- Your soul has incomprehensible potential powers.
- Your soul connects with your heart and mind. Soul can teach heart and mind. Your soul can transmit its great wisdom to your heart and mind.
- Your soul can connect with your Heaven's Team, which includes your spiritual guides, teachers, angels, and other enlightened masters in Heaven.
- Your soul stores an immense amount of information and messages. After you open your spiritual communication channels widely you will be able to access those messages and information anytime and anywhere.

- Your soul is constantly searching for knowledge. Just as your mind is always learning, so too is your soul. Your soul can learn from other souls, particularly from your spiritual fathers and mothers. Your soul has the potential to learn Divine and Tao wisdom and knowledge.

- Your soul can protect your life. "Outer" souls, including angels, saints, spiritual guides, enlightened teachers, and the Divine, can also protect your life. They can help you prevent sickness, change a serious accident to a minor one, or help you avoid an accident completely.

- Your soul can reward you as well as give you warnings. If your soul is happy with what you are doing, your soul can bless your journey. If your soul does not like what you are doing, it can make your life difficult. It could block your relationship or even make you sick.

- Your soul can predict your life. If you communicate with your soul, it can tell you what is in store for you.

- Souls follow spiritual laws and principles. Your mind may not be aware of this, but your soul absolutely follows spiritual laws.

- Many souls yearn to be enlightened. To reach ultimate soul enlightenment is to become totally quantum entangled with everyone and everything. It is to be connected with and serve everyone and everything. Our souls want to offer good service in the form of love, care, compassion, sincerity, generosity, and kindness. That is why more and more people are searching for soul secrets, wisdom, knowledge, and practices.

- Your soul is eternal.

Soul Power and Information: Positive and Negative

There are two kinds of soul power: positive and negative. Positive soul power comes from the positive information within us. Negative soul power comes from the negative information within us. Positive information brings about connection, coherence, health, rejuvenation, longevity, healing, enlightenment, success in relationships and finances, and more. Negative soul power brings sickness, depression, pain, fear, grief, suffering, challenges in relationships and finances, and all kinds of disasters and catastrophes.

An important insight and revelation came to us. The information described by Claude Shannon and other physicists is in fact negative information. Negative information relates to the unrelated possibilities, the disorder, the disconnection, and the uncertainty in a system. The measure of negative information is entropy.

There is another type of information: positive information. Positive information measures the order, connection, and certainty that a system has within itself and with others.

A good example of a system that has positive information is a crystal. Within a crystal, every molecule or ion is aligned in a predictable way. A crystal has order. This order is part of the crystal's positive information.

Another example of positive information is the life system. For life to occur, order and connection must be present. In a healthy body, every cell is connected with every other cell and behaves in an orderly way. Our bodies have positive information.

In mathematics, a fractal has order. Even with all its complexities, every point in the fractal can be predicted. This is positive information.

In general, symmetry is a kind of order and positive information.

Positive information also exists in two quantum systems that are quantum entangled with each other. Information about the state of one system gives us information about the other system. This kind of order and connection is positive information.

Most things have both order and disorder co-existing simultaneously. Liquid water has both order and disorder. Ice has more order than liquid water. When water boils, it changes into steam. Steam has more disorder than water. The change from ice to water or water to ice, liquid water to steam or steam to liquid water is called phase transition in physics. It is a transformation from a state with higher order to a state with less order or from a state with lower order to a state with more order. Phase transition is the transformation of information within a system.

Positive information is measured by negative entropy, also referred to as negatropy. Negatropy shows the amount of order within a system.

The energy related to negative information is heat. The energy associated with positive information is called free energy in thermodynamics, chemistry, and biology.

Everyone and everything could have both negative information and positive information in their vibrational field. Negative information usually appears as darkness in the vibrational field. It is impurity in our vibrational field. Purification is to remove negative information in our vibrational field. The purer we are, the less negative information and the more positive information we have in our vibrational field.

The Power of Positive Information and Negative Information

Positive information is the connection we have with others and the order within ourselves. The more connection and quantum entanglement we have with others, the more our actions can affect others. Therefore, the more positive information we have, the more shen power we have. Shen power includes soul power, heart power, and mind power.

The amount of positive information we have determines the quality and power of our souls, hearts, and minds. If we have more positive information, we will have more soul power, heart power, and mind power. We will be more powerful, influential, knowledgeable, and healthier. We will live longer. The more positive information an organization has, the more efficient, productive, harmonious, healthy, and influential it will be.

Negative information is the uncertainty and disorder within us and our disconnection from others. Negative information will cause us to have less impact, less knowledge, less abundance, less love, and less joy. It will reduce our shen power, including soul power, heart power, and mind power. If one has a lot of negative information, one will more likely be sick, decay, die, and face difficulties and challenges in relationships, career, finances, and every aspect of life.

Many people talk about positive or good energy. We are drawn to people with good energy. People with good energy are usually happier, healthier, and more successful. This positive or good energy is energy that carries positive information. Positive or good energy brings order, connection, coherence, rejuvenation, flourishing, harmony, love, joy, and longevity to a system. There is also negative or bad energy. Negative or bad energy is energy that carries negative information. This negative energy brings disharmony, suffering, pain, disorder, disasters, challenges, difficulties, accidents, disconnection, sickness, decay, and death to people and organizations.

Positive information is what gives our souls, hearts, and minds miraculous power. Current physics focuses mostly on studying negative information, which is entropy. An object with only negative information will not have soul power, heart power, or mind power. Partly due to this reason, current physics has not discovered the power of our souls, hearts, and minds.

The Purpose of Life is to Enhance Positive Information and Soul Power

If you observe the life systems within you and around you, you will find that life is a system that is capable of taking in energy and matter to maintain and enhance its positive information, which is the order and connection it has. The entire life system aims to enhance its positive information and create higher levels of order and connection.

For example, a plant takes in matter from the air, from the earth, and from waste discharged by other plants, animals, and humans. It absorbs light from the sun. It utilizes the energy of the light to turn the matter with lesser order, such as dirt, air, and waste, into matter with higher order, such as roots, stems, leaves, fruit, nuts, and more. The organic "food" produced by the plant has more positive information and higher free energy. It can nourish and rejuvenate human beings and animals, as well as plants. Human beings and animals can turn the organic food into something with even higher order, such as more sophisticated body systems, thoughts, music, books, and more.

We can see that a life system is an integrated whole. Its components work together to transform information, energy, and matter to higher levels of order and connection. It creates more and higher positive information.

A life system uses the mechanism of reproduction to pass on its information to the next generation. In this way, the descendants can continue learning from and processing the ancestor's past information so that they can create even higher levels of positive information.

Life is built on positive information, the negatropy. *The purpose of life is to accumulate and enhance positive information and soul power.* The more positive information a life system has, the healthier, wiser, and more powerful it becomes. To increase the positive information of a system or an organization is to reduce the risks and disasters it may face and to boost the effectiveness, influence, and service it can offer. To increase humanity's and Mother Earth's positive information will reduce wars, poverty, hunger, natural disasters, pollution, and all diseases.

In view of this, we propose a mathematical definition of life:

Mathematical Definition of Life

Life is a system that can maintain, increase, and develop positive information. Life brings connection, order, power, wisdom, health, rejuvenation, joy, love, and thriving.

The opposite of life is anti-life, which leads to decay, death, and more. We propose a mathematical definition of anti-life:

Mathematical Definition of Anti-life

Anti-life is a system that maintains, increases, and develops negative information. Anti-life causes separation, disconnection, disorder, decay, death, sickness, suffering, pain, sadness, anger, disharmony, disaster, challenges, difficulties, and more.

With the above definition, we can see that life has power over anti-life. Life always thrives. Anti-life will always fail. This is because by increasing positive information, life always brings thriving, health, and longevity whereas by decreasing positive information, anti-life causes decay, sickness, and disasters.

Everyone and everything has both life and anti-life within. Every moment, we are given the opportunity to bring life or anti-life to others and to ourselves. It is important to recognize this and pay attention to our thoughts and feelings, to what we hear and see, and to every action, behavior, and speech so that we bring life rather than anti-life to the world.

We can now also offer mathematical definitions for saint and anti-saint:

Mathematical Definition of Saint

Saint is the life that brings positive information to everyone and everything.

Mathematical Definition of Anti-saint

Anti-saint is the anti-life that brings negative information to everyone and everything.

Given these definitions, everyone and everything can become a saint; everyone and everything can become an anti-saint. To become a saint is to enhance one's own and others' power, health, longevity, influence, love, joy, wisdom, knowledge, prosperity, effectiveness, and more. To become an anti-saint is to bring more separation, disconnection, disorder, decay, death, sickness, suffering, pain, sadness, anger, disharmony, disaster, challenges, difficulties, and more to one's self and to everyone and everything.

There are four levels of saint: human saint, Mother Earth saint, Heaven saint, and Tao saint.

A human saint has reached the state of complete connection and oneness with all humanity. A human saint has removed all negative information relating to the human body and humanity. This state is one of returning from old age to the baby state, with total health and purity. In this state, a human saint has the ability to influence human affairs at a high level and bring love, peace, and harmony to humanity.

A Mother Earth saint has established a complete connection and reached the state of oneness with Mother Earth. In this state, a Mother Earth saint has the ability to impact affairs on Mother Earth, including natural events such as weather, and help bring love, peace, and harmony to Mother Earth.

A Heaven saint has formed a complete connection and reached oneness with Heaven, including all the planets, stars, galaxies, and universes. A Heaven saint has the ability to affect heavenly events.

A Tao saint is completely aligned in oneness with Tao, the source of all. A Tao saint has melded with Tao, thereby transcending all cycles of life and death, reaching the highest and final enlightenment, and attaining the ultimate freedom and bliss.

What Is the Divine?

The Divine is the universal vibrational field that contains and is connected with everyone and everything. The Divine has complete positive information. Because the Divine is connected with everyone and everything, the Divine can hear, feel, know, and make anything happen instantly. The Divine is omnipresent, omniscient, and omnipotent.

The purpose of life is to enhance positive information to reach the Divine and Tao.

Encountering the Divine and Tao is the most important awakening in our lives.

Feeling the Divine's and Tao's love nourishing, rejuvenating, and providing everything we need is the highest blessing.

Realizing the Divine and Tao are with us in everything, everywhere, and every moment is the greatest wisdom.

Knowing the Divine and Tao is the supreme knowledge.

Experiencing the Divine and Tao in everyone and everything is the loftiest endeavor.

Reaching the Divine and Tao is the highest purpose of life.

Deepening our connection with the Divine and Tao is the highest practice.

Being one with the Divine and Tao is the greatest power.

Union with the Divine and Tao is the ultimate bliss.

The goal for all beings is to reach the Divine and Tao.

How to Develop the Power of Soul

To develop and enhance our soul power is to increase the positive information within us, which is to boost order and connection within ourselves and with others.

There are many ways to raise our positive information. Several important ones include:

- Chant sacred mantras. A sacred mantra is words, phrases, or sounds that carry positive information. By chanting (repetitively saying or singing) a sacred mantra, its vibrational field with high-level positive information can enhance the positive information in our vibrational field.

Chanting sacred mantras is a sacred spiritual practice in many spiritual traditions. One of the sacred practices in Buddhism is to chant *A Mi Tuo Fo*, the name of the Buddha of Infinite Light.

Hai Xian became a Buddhist monk when he was nineteen years old. He was illiterate. His teacher only taught him to chant *A Mi Tuo Fo*. That is what he did for ninety-two years until he transitioned at the age of one hundred twelve. He knew the exact day he would leave Mother Earth. Until that day, he continued to work. He left instructions not to be buried, but for his body to be kept in a bucket after his death. Six years after his death, people opened the bucket and were amazed to find that his remains were intact. His body was as fresh as if he were still alive. Why didn't his body decay even after his death? His chanting of *A Mi Tuo Fo* transformed his body into matter with complete positive information. Matter with complete positive information does not decay. Hai Xian demonstrated the power of chanting the sacred mantra *A Mi Tuo Fo*.

- Read and write sacred texts with positive information. Listen to sacred sounds with positive information. Look at sacred objects with positive information. Be in the presence of spiritual masters with positive information.

- Offer service to others. To serve is to make others healthier, happier, wiser, or more peaceful. It is one of the most effective ways to build positive information, connection, and order in the world.

- Meditate. To meditate is to connect with Tao, allowing Tao's vibrational field to transform our vibrational field.

- Develop positive qualities such as love, forgiveness, compassion, light, humility, harmony, gratitude, service, and enlightenment.

- Receive and practice with Tao downloads/transmissions from an authorized Tao servant. Tao downloads carry Tao vibration that can transform your vibrational field to a Tao vibrational field. We will explain this further at the end of this chapter and actually offer you, dear reader, Tao downloads in chapter eight. The power of this spiritual practice is beyond logic and comprehension.

- Trace and write Tao Calligraphy. We introduce this powerful oneness treasure and practice in chapter nine.

How to Use the Power of Soul

Your soul has miraculous power to gain wisdom and receive information, to communicate, and to heal, transform, and uplift every aspect of your life. We are honored and delighted to share some techniques for applying your soul power. These techniques include the Say Hello Technique, Soul Orders, soul communication, and soul downloads and transmissions. They can literally create miracles in your life.

As you have learned, your soul is information. You can simply ask or communicate with your own soul and other souls to help you accomplish certain goals. This is how prayer, the Say Hello Technique, Soul Orders, soul communication, soul healing, soul rejuvenation, soul prevention of sickness, soul boosting of energy, soul transformation of relationships and finances, soul marketing, and soul conferencing work. Information can also be downloaded and transmitted to your soul. This is how soul downloads and transmissions work. For some people, these techniques may be too simple and miraculous to believe. However, Tao Science tells us that these techniques have a scientific basis, although they may appear to be "magic."

I have taught in great detail in my book, *The Power of Soul*, and other books how to use these techniques to heal and transform health, relationships, finances, business, intelligence, spiritual channels, and more. These techniques have created millions of miracles and have transformed many people's lives. Here we will only briefly describe these techniques. Please consult my other books (see page 253 for a selected list) to learn much more.

Say Hello Technique

The simple Say Hello Technique is based on the important wisdom that we can directly ask souls to bless our lives. You can ask your own soul and any of your inner souls, including the souls of your systems, organs, cells, cell units, DNA, RNA, or parts of your body, to bless your life and deliver their soul wisdom and knowledge to you. You can also ask all kinds of outer souls, including the souls of Jesus, Buddha, all kinds of spiritual fathers and mothers, holy beings, angels, oceans, rivers, mountains, Mother Earth, the sun, the moon, stars, galaxies, and universes, to bless your life and deliver their intelligence, wisdom, and knowledge to you.

Say *hello* means to invoke and ask the inner and outer souls of your choice for healing, blessing, wisdom, and knowledge. The Say Hello Technique involves a five-step formula:

Step 1. Address (say *hello* to) the souls:

> *Dear shen qi jing of* ________ (name the inner souls and outer souls you wish to invoke),

Step 2. Honor the souls:

> *I love, honor, and appreciate you.*

Step 3. Make an affirmation:

> *You have the power to* ________ (state your request here).

Step 4. Make your request for healing, blessing, wisdom, knowledge, or other:

> *Please* ________ (repeat your request).
>
> *Do a good job.*

Step 5. Express gratitude:

> *Thank you. Thank you. Thank you.*

Let's look at an example where we invoke some outer souls:

> *Dear shen qi jing of Tao, dear shen qi jing of the Divine, dear shen qi jing of all my spiritual fathers and mothers,*
>
> *I love, honor, and appreciate you all.*
>
> *Your love and power can transform every aspect of my life.*

Please give me a healing and blessing for ________ (state your request).

I am deeply grateful.

Thank you. Thank you. Thank you.

Every soul has its power to heal and bless. The Say Hello Technique is to invoke souls to heal, bless, and create. Tens of thousands of heart-touching healing successes and miracles from all over the world have been created by applying the Say Hello Technique.

Now think of an area in your life that needs healing. It could be your physical body, emotional or mental issues, or challenges with your relationships or finances. You can use the Say Hello Technique to help you solve it.

Let's do it now. Suppose you have an issue with your colleagues. You would like to improve your relationships with them. Here's how you can use the Say Hello Technique to help:

Dear shen qi jing of my colleagues and me,

Dear shen qi jing of the relationships between my colleagues and me,

I love you, honor you, and appreciate you.

You have the power to heal my relationships with all of you.

Please bless us to have better relationships with each other from now on.

Thank you. Thank you. Thank you.

Dear shen qi jing of Jesus, Mother Mary, and Buddha,

Dear the moon, the sun, and countless planets, stars, galaxies, and universes,

You have the power to transform my relationships
with my colleagues.
Please give a healing and blessing to the
relationships between my colleagues and me.
I am extremely grateful.
Thank you. Thank you. Thank you.

You can enhance the power of the Say Hello Technique by combining the Soul Power with Body Power, Mind Power, and Sound Power. Together, these are called the Four Power Techniques.

Body Power is to use hand and body positions to heal. *Where you put your hands is where energy and healing go.* For relationships, you can put your hands on your heart. In traditional Chinese medicine wisdom, to release anger, you can put your hands on your liver. To release anxiety or depression, you can put your hands on your heart. To release worry, you can put your hands on your spleen. To release your sadness and grief, you can put your hands on your lungs. To release fear, you can put your hands on your kidneys.

Mind Power is to use the mind or consciousness for healing. *Where you put your mind, using imagination and creative visualization, is where you receive benefits for healing.* The more light in the vibrational field of the relationship, the healthier the relationship is. Now imagine and visualize the souls of the relationships between your colleagues and you to be golden, rainbow, or crystal light balls, radiating brighter and brighter with more and more light.

Sound Power is to chant sacred mantras. A mantra is a sacred word, text, or sound that carries positive information. Chanting a mantra brings its vibrational field to you. *What you chant is what you become.* Chanting a sacred mantra can enhance the positive information within your soul, heart, mind, and body (energy and matter). It can also help enhance positive information in the world. Chant sacred mantras with all your soul, heart, mind, and body. Chant as often and as long as possible. It can help heal, transform, uplift, and enlighten your soul, heart, mind, body, and every aspect of your life.

Love melts all blockages and transforms all life. Forgiveness brings inner peace and inner joy. Light heals and prevents all sickness. Now let's chant:

Love
Love
Love
Love
Love
Love
Love
Love
Love

Forgiveness
Forgiveness
Forgiveness
Forgiveness
Forgiveness
Forgiveness
Forgiveness
Forgiveness
Forgiveness

Light
Light
Light
Light
Light
Light
Light
Light
Light

Now chant:

Loving relationships
Loving relationships
Loving relationships
Loving relationships
Loving relationships
Loving relationships
Loving relationships
Loving relationships
Loving relationships ...

Chant like this for five to ten minutes a few times a day. Chant especially when you feel the challenges in your relationships. If you continue to practice daily, you will start to notice that the relationships between your colleagues and you are improving—perhaps miraculously!

You can apply the Say Hello Technique and Four Power Techniques to help heal, transform, and bless every aspect of your life and the lives of others. It is so easy and so powerful. It can create many miracles for you and others.

Now think of another area in your life and apply these techniques again. Experience the power of the soul.

Soul Orders

A Soul Order is exactly what it sounds like: an order given by a soul to do something that is good service, such as healing, preventing sickness, rejuvenating, transforming life, and enlightening life. A Soul Order can be given to yourself by your body soul, or by any of your inner souls, including the souls of your systems, organs, cells, cell units, DNA, and RNA.

A Soul Order can be given for self-healing. For example, if you have back pain, the back has a soul. The soul of your back can give an order to heal your back. Here is how you can do it:

Dear soul of my back,
I love you, honor you, and appreciate you.
You have the power to send an order to heal my back.
Please send an order to heal my back.
The order is: Soul of my back orders my back to heal.

Then, activate the order by chanting repeatedly:

Soul of my back orders my back to heal.
Soul of my back orders my back to heal.
Soul of my back orders my back to heal.
Soul of my back orders my back to heal. …

Now let's add Body Power and Mind Power. Put your hands on your back. Imagine golden, rainbow, or crystal light radiating more and more brilliantly in your back. The light becomes brighter and brighter. Imagine a sun inside your back radiating blinding light. Continue to chant as you do this. You may chant aloud or silently.

Repeat this order for at least three minutes. There is no time limit. The longer you chant, the better. If the pain continues, do this several times a day. After a short while, you may find that your back pain has significantly improved.

You can apply Soul Orders in any area of your life. Now think of another area of your life that you would like to improve. Give a Soul Order.

You can also give a Soul Order before you go to sleep. In this way, your soul can work on your issue while you sleep. Try it and see how it improves your life.

Soul communication

Souls carry information, messages, wisdom, and knowledge beyond comprehension. You can receive information, messages, wisdom, knowledge, and insights about the past, present, or future, about a place, a thing, or an event through soul communication.

In the first two books of my Soul Power Series, *Soul Wisdom*[5] and *Soul Communication*,[6] you can learn how to open your spiritual communication channels to communicate not only with your own soul, but also with the souls of Tao, the Divine, Buddha, Jesus, Mother Mary, other saints, Mother Earth, plants, animals, rivers, oceans, mountains, and countless

---

[5] *Soul Wisdom: Practical Soul Treasures to Transform Your Life.* New York/Toronto: Atria Books/Heaven's Library Publication Corp., 2008.

[6] *Soul Communication: Opening Your Spiritual Channels for Success and Fulfillment.* New York/Toronto: Atria Books/Heaven's Library Publication Corp., 2008.

planets, stars, galaxies, and universes. Soul communication, which is spiritual communication, can empower and uplift your life beyond words.

Soul communication is not limited by space and time. It does not have to be done in person. For example, suppose you have an issue with your neighbor. They may be angry and unforgiving. It may not be easy—maybe even not possible—to speak in person. What you can do is speak with your neighbor's soul and resolve your issue through soul communication. You can do it like this:

Dear soul of my neighbor,
I love you, honor you, and appreciate you.
Please forgive me for anything and everything I have ever done wrong to you in all lifetimes.
I completely and unconditionally forgive you for everything you have ever done wrong to me.
Let's be good neighbors from now on.
Let's live in love, peace, and harmony.
Thank you. Thank you. Thank you.

Then, chant repeatedly for a few minutes:

Please forgive me.
I forgive you.
Love, peace, and harmony. ...

There is no time limit. The longer you chant, the better. Chant with sincerity, intention, and focus. Chant with and from your heart.

Do this simple soul communication practice every day. You may quickly notice an improvement in your relationship with your neighbor.

You can use soul communication for business, finances, and marketing, for health and rejuvenation, for personal growth and relationships, for manifestation and creation, and more. You can hold a soul conference by inviting other souls to get together for some positive purpose. You can broadcast information, announcements, and invitations to souls.

You can also use soul communication to receive wisdom, knowledge, and guidance for your life journey. You can receive breakthrough ideas and information. Indeed, the breakthrough concepts of Tao Science, as they appear in this book and in various popular and scholarly articles, were received through soul communication. We can say with total confidence that Tao Science will continue to develop through soul communication by us and others.

In fact, you, we, and all souls are doing soul communication every moment. We can all enhance our abilities to do soul communication and use it in every area of our lives in more effective and positive ways. This can enhance our lives beyond our comprehension.

To enhance our ability to do soul communication, we need to open our soul communication channels. As we shared in the introduction, major soul communication channels include the Soul Language Channel, the Direct Soul Communication Channel, the Third Eye Channel, and the Direct Knowing Channel. There are many ways to develop your soul communication channels. As a brief introduction, here is one simple practice you can do to open or further open and develop your soul communication channels.

Practice to open your soul communication channels

Sit up straight. Put your left palm over your Message Center[7] (heart chakra). Put your right hand in the prayer position with the fingers pointing upward. This hand position (Body Power) is a special signal to connect your soul with the Divine and the Soul World. We call it the Soul Light Era Prayer Position. It focuses on your Message Center because that is the key energy center for the potential power of your soul.

Now visualize golden light shining and radiating in your Message Center.

Apply the Say Hello Technique:

> *Dear shen qi jing of my soul communication channels,*
>
> *I love you, honor you, and appreciate you.*
>
> *You have the power to further open and develop yourselves.*
>
> *Do a good job!*
>
> *Thank you. Thank you. Thank you.*

Next, we will use the sacred mantra San San Jiu Liu Ba Yao Wu, which is Chinese for the number sequence 3396815. This sacred mantra, pronounced *sahn sahn jeo leo bah yow woo*, is a divine code for opening soul communication channels and unlocking the potential power of your soul. My (Master Sha's)

---

[7] The Message Center or heart chakra is one of the body's key energy centers. It is fist-sized and is located in the center of the chest, behind the sternum at the level of the nipples. It is the center for receiving information and messages, and so it is the soul communication center.

teacher, Master Zhi Chen Guo, received this sacred mantra in 1974 during his meditation.

Invoke this sacred mantra:

> *Dear shen qi jing of the sacred mantra* San San Jiu Liu Ba Yao Wu,
>
> *I love you, honor you, and appreciate you.*
>
> *You can help open and develop my soul communication channels.*
>
> *I am truly grateful.*
>
> *Thank you. Thank you. Thank you.*

Now chant *3396815* repeatedly as fast as you can. Do it now:

> *San San Jiu Liu Ba Yao Wu*
>
> *San San Jiu Liu Ba Yao Wu*
>
> *San San Jiu Liu Ba Yao Wu*
>
> *San San Jiu Liu Ba Yao Wu ...*

Chant faster and faster. Let go of any conscious intent to pronounce the individual numbers clearly. As you chant faster and faster—as fast as you can—suddenly, a special voice you may not have heard before could flow out. This special voice is your soul's voice speaking Soul Language. Some people can bring out their Soul Language easily. Others need to do the above practice longer. You can find more practices—and blessings!—in my books, *Soul Wisdom* and *Soul Communication*, for opening your soul communication channels further.

When you bring out your Soul Language, you are bringing out the information, messages, wisdom, knowledge, and power of your soul. To receive this information, messages, wisdom, and knowledge, you need to be able to translate your Soul Language into your human language. Let's do a practice to help you develop your Soul Language translation abilities. First, resume your Soul Light Era Prayer Position (page 80) with your hands. Focus your mind on your Message Center. Apply Soul Power with the Say Hello Technique:

Dear shen qi jing of my Soul Language translation abilities,

I love you, honor you, and appreciate you.

You have the power to translate my Soul Language into English.

Do a good job!

Thank you. Thank you. Thank you.

Then chant:

San San Jiu Liu Ba Yao Wu
San San Jiu Liu Ba Yao Wu
San San Jiu Liu Ba Yao Wu
San San Jiu Liu Ba Yao Wu …

Chant *San San Jiu Liu Ba Yao Wu* until your Soul Language comes out. Speak your Soul Language for about one minute. Then, open your mouth and speak in English without using your mind. What you speak will be the translation of your Soul Language. Some people can bring out this ability quickly; others may take weeks, even months of practice. Be persistent. Speak your Soul Language often. The benefits are beyond imagination.

Soul downloads and transmissions

Because a soul is quantum information, you can download and transmit information to your soul. You can use the Say

Hello Technique and Soul Orders to download and transmit information to your soul. One's soul can only access the information that it is quantum entangled or connected with. The more positive information one has, the more information one can download. Therefore, a high-level spiritual master can download otherwise inaccessible information to your soul and can offer you huge blessings and empowerments for health, rejuvenation, wisdom, knowledge, intelligence, relationships, finances, spiritual communication channels, and more.

For example, suppose you want to learn hula, the sacred beautiful ancient Hawaiian dance. You can connect with the soul of hula and ask it to download and transmit the wisdom, knowledge, and practice of hula to you. Then, when you start to learn hula from a physical teacher, you could find learning easy and quick. Or, suppose you want to write a book. Decide the title, and then ask the soul of the book to download the wisdom, knowledge, structure, language, tone, look, feel, and more to you. Actually writing the book could then require much less effort.

In closing, we want to emphasize two important things you need to know about soul downloads and transmissions. First, the information that you can download and transmit, or that is available for you to receive, is determined by your soul standing. Souls have different levels. Soul standing is determined by the amount of quantum entanglement the soul has. The more quantum entanglement your soul has, the higher your soul's standing. The higher your soul's standing, the more information you can download, transmit, and receive.

Second, it is important to exercise caution when you choose to receive a download or transmission from a teacher, a healer, or a spiritual master. Not everyone and everything has information that is of the highest benefit to you. You and others may have harmful information. To download, transmit, or receive such negative information will not benefit you or others.

Because of these two reasons, we need to be careful about the downloads and transmissions we ask for and how and, especially, from whom we receive them.

I (Master Sha) am honored and delighted to offer you a special and powerful soul transmission as a gift in chapter eight.

Soul power is the power of the twenty-first century. It is the miraculous power we all have. Knowing, developing, and using our soul power will take each one of us and all humanity to a higher level of existence. The significance of the soul's wisdom and power is beyond our comprehension. We wish that you will develop your soul power and fulfill your highest life purpose.

5

Power of Heart and Mind

IN TAO SCIENCE, the heart is the receiver of information and the mind is the processor of information. Heart and mind determine the nature of the physical reality we experience and of the life we have. Understanding and using the power of heart and mind is critical for health, relationships, finances, happiness, and success in every aspect of our lives. This profound wisdom has been known for millennia by sages, saints, buddhas, and spiritual masters. Unfortunately, many people have not reached a deep realization of the power of the heart and mind. Therefore, many people are not able to use the power of their hearts and minds in a positive way. This has caused much suffering.

Tao Science gives us scientific insights and understanding of the power of the soul, heart, and mind. We believe that true understanding of the power of the soul, heart, and mind can be an important turning point not only for your life, but for humanity. It can release much suffering. It can empower us to become a powerful creator and manifestor of a life we really want.

Are You an Observer or a Creator?

Your soul is the content of its information. It contains many possibilities and potentialities. Your vibrational field contains countless vibrations and states. Yet, your physical reality usually consists of certain specific outcomes. How does a single concrete reality emerge from the many possibilities?

The measurement problem in quantum physics involves the question of why our world appears to be definite while the underlying quantum nature is the superimposition of many possible vibrational states. The main issue in the measurement problem is how observed reality is manifested from the vibrational field, which contains many possible states described by the wave function.

What you will find in quantum physics is truly breathtaking. As it turns out, you are not a passive observer of a quantum phenomenon, you are creating the phenomenon you observe!

How does this happen? Let us explain the manifestation process in Tao Science.

The first thing you need to know is that to observe a quantum phenomenon, you must receive vibrations from what is being observed.

In quantum physics, detectors are used to receive vibrations. A detector is an instrument specially designed to absorb vibrations and show visible and measurable changes as a result. For instance, a camera is a detector. Photographic film or digital detectors are used to detect visible images, X-rays, and more. A radio is a detector that receives radio waves and broadcasts them. A television is a detector. It receives waves from television stations and shows the programs through sound and images. Our eyes, ears, nose, skin, and tongue are all detectors. Our eyes detect light. Our ears detect sound. Our nose detects smell. Our skin detects temperature. Our tongue detects taste. Our heart is a detector.

The second thing we need to know about quantum phenomena is that the kinds of detectors we use and where we place the detectors determine what we observe. Different detectors exhibit different phenomena because they absorb different vibrations. For instance, if you use a normal camera, you will see images of visible light. If you use a special camera that can receive infrared light, you will catch images of infrared light. We have all had the experience that where we place our camera and the angle at which we direct our camera greatly affect the resulting photograph.

In summary, the kinds of detectors we use and where and how we place the detectors determine the phenomena we observe. Because of this, quantum phenomena depend on us, the observers. We are in fact actively participating in creating the observed phenomena. Quantum phenomena are intrinsically subjective, meaning they depend on the observer.

Our detectors include our hearts and minds. Quantum physics reveals to us a profound and most empowering truth, one which Buddha, Jesus, and many other spiritual teachers and masters have taught us:

Buddha taught "Xiang You Xin Sheng." Xiang means *image* or *phenomenon*. You means *comes from*. Xin means *heart*. Sheng means *create*. Xiang You Xin Sheng literally means *all the images we see and all the phenomena we experience come from our heart*.

In the Bible, Proverbs 4:23, it is said: "Above all else, guard your heart, for everything you do flows from it."

Many other spiritual traditions offer similar wisdom and guidance.

We create our own reality. Our own hearts and minds manifest our physical reality from our souls.

One of the famous debates in quantum physics is whether an electron exists if you are not observing it. We can answer this question easily in Tao Science. If you are not observing it, an electron exists in possibility but it does not exist in our reality. Our action of observation is actually manifesting the electron into our reality.

Therefore, everyone's reality can be different. We all have this experience in life. Different people have different experiences, ideas, and feelings about the same event. In scientific studies, researchers try to control all the detectors so that the experiment is repeatable.

We must completely understand that our physical eyes and modern technology are limited. So much of what exists is invisible to our physical eyes and our most powerful instruments. Since we can only observe certain things, we can only manifest certain things.

Think of your life as a movie. Quantum physics tells us that what is in your life movie is actually manifested by your own actions, thoughts, intentions, speech, hearing, smelling, feeling, movement, and more. You are not only the observer and actor, but also the producer and director of your own life movie. You are the creator of your life!

Power of Heart

"Heart" includes the physical heart and the spiritual heart. Your spiritual heart is the receiver of information. Your spiritual heart corresponds to the detectors used for observation in quantum physics. It includes your thinking, feelings, seeing, hearing, smelling, tasting, speaking, emotions, and more. The one-sentence secret about the power of heart is:

What we receive in our hearts determines what we manifest.

Everything in your life, including your body, health, relationships, career, intelligence, family, and every aspect of your life, comes from your heart. Your heart activities, which include your thinking, intentions, speech, hearing, smelling, feeling, movement, tasting, and more, decide what happens in your life.

Your heart plays a crucial and determining role in the creation and manifestation of your reality. What kind of heart you have and where you put your heart determine your life. This process is similar to television. The program you watch on television is determined by the station you choose. If you set your heart on love, the physical life and reality you experience will be full of love. If you put your heart on sorrow, your reality will also be sad.

Heart power includes positive heart power and negative heart power. Positive heart power is heart action that increases your soul power. To increase your soul power is to increase the order and connection you have. Positive heart power includes love, forgiveness, compassion, light, humility, harmony, abundance, gratitude, service, and enlightenment. Positive heart power empowers and uplifts you. It helps you manifest a life full of love, abundance, joy, harmony, and peace.

Negative heart power is heart action that decreases your soul power. It reduces the order within you and your connection with others. Negative heart power includes selfishness, competition, jealousy, anger, sadness, grief, anxiety, depression, fear, pride, hatred, feelings of inferiority or superiority, and more. Negative heart power disables and limits you. It will manifest a life with more suffering.

It is of utmost importance that you be aware of the kind of heart power you are applying in every moment. Apply positive heart power to manifest a life full of love, joy, abundance, peace, beauty, and wisdom. Avoid negative heart power to avoid a life filled with challenges, difficulties, anxieties, pain, and suffering.

Many people put great attention on what they eat and drink. Some people devote great effort to money and wealth. Some care a lot about their social status. What we eat and drink, how much money we have, and our social status are important. However, the determining factor for the kind of life we have is what is going on in our hearts. What our hearts receive determines our health, relationships, career, finances, longevity, and every aspect of our lives. Being mindful and careful of what is going on in our hearts is crucial for creating the life we want.

Power of Mind

The mind processes information received by the heart. The mind has amazing abilities to facilitate the process of manifestation. The abilities of the mind include imagination, creative thinking, visualization, discipline, judgment, using tools, analyzing, synthesizing, integrating, and more. Many people have explored the power of the mind.

Humanity has invented machines and computers to expand our ability to process information. However, we have not used our own mind power enough. Most of us use less than five percent of the potential of our brains. Each of our cells, tissues, organs, and systems is constantly processing information as well. Our own capacity to process information is beyond our comprehension.

The mind directs where one's energy and matter go. The one-sentence secret about mind power is:

The mind determines which of the heart's desires are manifested and how much and how quickly they are manifested.

When you use your mind power correctly, you can manifest your heart's desires quickly and in the most magnificent way.

Mind power can also be divided into positive and negative. Positive mind power expedites, expands, and enhances the fulfillment of the heart's desires. Positive mind power also increases one's positive soul power.

Negative mind power delays, prevents, and stops the heart's desires from being fulfilled, and also decreases one's positive soul power. Negative mind power includes negative mind-sets, negative beliefs, negative attitudes, ego, attachments, and more. It is essential to be aware of and remove these aspects of negative mind power if you want to lead a happy and successful life.

It is important to realize that the mind's purpose is to serve the heart and soul. It is essential to listen to your heart and soul, and then use your mind to execute and follow through on your heart's and soul's directions. Many people allow their minds to take the lead instead. This has caused many issues, including suffering, diseases, depression, anxiety, relationship challenges, financial difficulties, and more.

Soul is the boss. Heart should listen to the soul. Mind should follow the heart. To follow this process is to follow nature's way. It is to follow Tao. It is important to direct, train, and use your mind to align with your heart and soul. Then, your mind can truly serve your heart and soul. Only then can you accomplish the highest goals of your life journey and your soul journey.

Resonance: How the Heart Receives

How does the heart receive information from the soul? The heart receives information through a process called resonance.

Every quantum vibration shares an important quality. When a quantum vibration is absorbed or emitted, it will be absorbed or emitted completely. It cannot be absorbed or emitted partially. In this sense, a quantum vibration behaves like a particle. This is often called the particle nature of quantum vibration.

In quantum field theory, this phenomenon is described and explained mathematically through resonance. The absorption of a quantum wave can only happen through resonance. Resonance is a property between an object and a quantum wave. Resonance occurs when an object has two states and the energy difference between these two states is equal to the energy of the quantum wave. Figure 2 below illustrates how resonance can happen.

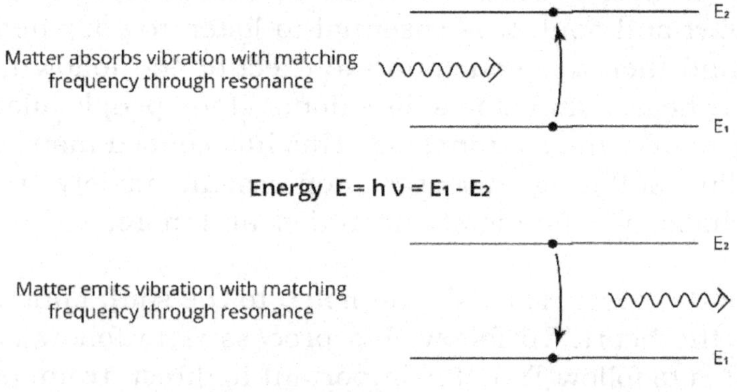

Figure 2. Absorption and emission of light through resonance

The vibration with frequency v carries the energy hv. Here h is the Planck constant. When the energy of an incoming vibration is equal to the difference of the two energy states within the object ($E_1 - E_2$), the object can absorb the incoming vibration. If this condition is not met, the vibration cannot be absorbed. In the same way, the object can emit a vibration with the frequency $h\nu = E_1 - E_2$.

You can observe the resonance phenomenon in musical instruments. A musical string can start to vibrate and produce sound through resonance with vibrations in the air, without being struck by an object or plucked by a person. This resonance phenomenon is caused by the existence of vibrations in the air with the same frequencies as those the string can produce. The string resonates with the vibrations of these frequencies, vibrates at the same frequencies, and produces sound.

Everyone and everything is a resonant system. Everyone and everything can absorb vibrations with specific frequencies from the environment through resonance. This tells us that the heart receives the vibrations with which it can resonate. This is how the heart receives information.

Through resonance, we communicate with each other. Through resonance, we receive information, energy, and matter. If we cannot resonate with something, then we cannot receive its information, energy, and matter. Similar things have vibrations with similar frequencies. Things that are similar more easily resonate with each other. In nature, similar things tend to gather together. This is because they can exchange information, energy, and matter more easily.

In Tao Science, we can apply Tao Source shen qi jing to purify a human's shen qi jing—and the shen qi jing of everyone and everything—in order to increase the amount of resonance. This will increase one's ability to receive information, energy, and matter in order to enhance health, relationships, finances, intelligence, spiritual abilities, enlightenment, and every aspect of life. This is Tao Source creation, wisdom, practice, and power.

How to Develop the Power of Heart

The power of your heart is determined by its potential for resonance. The more resonance your heart can have, the greater its power. To develop and enhance the power of your heart is to increase the totality of vibrations it can resonate with. The more vibrations your heart can resonate with, the more information, energy, and matter it can receive. In turn, you will receive and be able to use more wisdom, knowledge, energy, and matter.

Enhancing heart power is crucial for boosting soul power. When your heart has greater resonance, it can establish more connection with others. This positive heart activity can greatly enhance your positive information, and thus increase your soul power.

Enhancing heart power is also critical for increasing mind power. With greater resonance, your heart can make more information available for your mind to process and more energy and matter to direct. This will greatly enhance your mind power.

Cultivating greatest love, forgiveness, compassion, light, humility, harmony, flourishing, gratitude, service, and enlightenment are some of the best ways to increase heart power.

How to Cultivate the Power of Mind

Mind power is measured by how much information our mind can process in a given period of time and how much energy it can direct. The more information it can process and the more energy it can direct in a shorter period of time, the more powerful the mind is.

To cultivate the power of your mind, one of the most important methods is to remove mental blockages, such as ego, negative mind-sets, negative attitudes, negative belief systems, attachments, and more. Establishing peace and humility in your mind can greatly enhance your mind power.

How to Use the Power of Heart

The power of the heart is to feel, experience, receive, and manifest. The more deeply, strongly, and frequently you feel, think, see, hear, smell, and taste the things you want, the more quickly you can receive and manifest them. This is the power of heart.

To have deep heart-felt realizations, sincere intentions, strong messages, clear spiritual images, deep feelings, positive emotions, seeing, hearing, speaking, thinking, tasting, smelling, and more is to use the power of heart for creation and manifestation.

Is It Just Coincidence?

We all experience coincidences in our daily lives. We have all encountered totally unrelated events or events that are unlikely to occur together, but come about in a meaningful way. The renowned Swiss psychologist, Carl G. Jung, coined the word *synchronicity* in the 1920s to describe these phenomena.

Are coincidence and synchronicity really "just" coincidence? Or is there something more behind these everyday events?

If your spiritual communication channels are open, you will see there is no coincidence. There is even a profound reason why a specific stranger is sitting next to you on the airplane. Coincidences reveal the deeper meaning, connections, and relationships between souls.

Synchronicity and coincidence are due to the quantum entanglement of vibrations among different people and things. If some of the quantum waves within us are quantum entangled with some of yours, whatever we think, feel, and do can instantly affect you. Whatever you think, feel, and do can also instantly affect us.

What causes the quantum entanglement of these vibrations? There are many ways to create quantum entanglement among different vibrations. Normally, quantum entanglement is due to the fact that the quantum-entangled vibrations are created from the same source or through the same process, and are thus correlated or connected with each other. For example, one light vibration is split into one electron and one positron. A positron is a particle that has the same mass and spin as an electron, but opposite charge. The electron and the positron that are created from the same light wave are quantum entangled with each other.

Coincidence and synchronicity reveal an intrinsic connection among beings. Divination systems use synchronicity and coincidence to obtain information about the past, present, and future. For example, a tarot card reader uses the "coincidence" of the specific cards that are revealed and the order in which they are revealed to discover information for people. A shaman looks for "signs," which are coincidences, to predict what is going to happen. Understanding coincidence and synchronicity can help one understand the present better. It can also explain what has happened in the past and predict what will happen in the future.

Some people think that coincidence and synchronicity violate the law of cause and effect. They haven't realized that coincidence and synchronicity are caused by the creation of the involved beings from the same source. Thus, they are due to cause and effect. Understanding how coincidence and synchronicity weave through your lives and create your reality will empower you to become a powerful and miraculous creator and manifestor.

Many of us have had the experience that after having a deep-hearted realization, we suddenly notice that many coincidences or accidents occur that support our heart-felt truth. For example, one day, our friend Mimi had a heart-opening experience and came to a profound realization that love is all that is in the world. Next day, she went to town. She found that people treated her so lovingly. A waitress gave her an extra dish and dessert. Someone else gave her money. Another person wanted to give her a well-paying job. These are obviously not coincidences. Her own heart-felt realization and heart activities initiated these loving actions towards her. In more scientific terms, her heart emitted and received "lovingness" and thus manifested a loving physical reality. Her own love intention transformed her vibrational field of quantum entanglement. For a normal being, it could be viewed as coincidence and synchronicity, but it is not.

Coincidence and synchronicity are caused by our deep soul connections and our own heart activities. You may have had the experience that, one day, you suddenly realize you must do an important task. Suddenly, you find people appearing in your life to help you with your project. This is because some of us were assigned together to accomplish a task larger than any of us individually. When your heart receives the message to step into your role to fulfill your task, the people who were assigned to the same task will be activated to come into your life so that you can all accomplish the task together.

Paying attention to coincidence and synchronicity will empower you in a profound way. This holds the key to the next stage of human evolution.

In Tao Science, you will learn the wisdom and experience more coincidence and synchronicity to enhance every aspect of your life, including health, relationships, finances, spiritual channels, success, enlightenment, and more.

How to Use the Power of Mind

The purpose of mind power is to enhance soul power and heart power. It is to facilitate the manifestation of the soul information by the heart.

To develop your soul is to bring more light and more positive information to your vibrational field. When your spiritual channels are open, you can see that negative information appears as darkness in your vibrational field.

Creative visualization of light is one of the best ways to use your mind power.

Visualizing light in your vibrational field will help bring more light to your soul and thereby allow your heart to receive more light. It will enhance the power of your soul and heart.

Visualizing light inside your body will help you heal and rejuvenate your body faster. Visualizing light in one of your relationships will help improve that relationship. Visualizing light in your business and career will bring more success to your business and career.

Learn this sacred practice to use your mind power to heal, transform, and uplift every aspect of your life.

6

Power of Energy and Matter

WE ARE FAMILIAR with our physical reality. Matter makes up our physical reality. We also know energy is essential for us to get things done. Many people strive to accumulate energy and matter to have a comfortable life.

In Tao Science, energy and matter have deeper meanings and functions. Energy is the actioner to enable action. Matter is the transformer to achieve what we come to this world to accomplish. Energy empowers us to take action. Matter has the power to transform our souls, hearts, minds, and every aspect of our lives.

The purpose and function of life is to enhance connection and order, which is positive information. Increasing the positive information within us will boost the power of our souls, hearts, and minds. It will uplift every aspect of our lives. Increasing the positive information in others will boost the power of their souls, hearts, and minds. It will uplift every aspect of their lives. The purpose of life is to accumulate positive information and transform negative information. Our energy and matter—our actioner and transformer—should serve this purpose. Attaching to physical reality and materialistic gain will not only limit the success and potential of our lives, it will actually waste our precious lives by preventing us from achieving our true life purpose.

Energy and matter can be divided between positive and negative. Positive energy and positive matter can lead to love, joy, wisdom, health, longevity, growth, upliftment, abundance, and enlightenment. Negative energy and negative matter can lead to sickness, pain, suffering, disasters, challenges, and difficulties in relationships, finances, and more. It is crucial to know what kind of energy and matter we are using in our lives.

Positive Energy and Negative Energy

In physics, energy can be both positive and negative. Positive energy makes actions, such as moving objects, possible. Negative energy stops actions and can get things stuck. For example, adding gasoline to your car will give your car positive energy, which will make your car run longer. Gravity, being the force of attraction, can generate negative energy, which keeps your car earthbound. Your car cannot fly into space because of the negative energy created by gravity. It takes positive energy, for example, in the form of a rocket, to overcome negative energy and propel your car into space.

In Tao Science, energy is the actioner that makes actions happen. There are also two kinds of action: positive action and negative action. Positive action is action that increases positive information. Negative action is action that increases negative information.

In addition to the normal positive energy and negative energy in physics, there exists another type of positive energy and negative energy in Tao Science, namely, energy that enables positive action and energy that enables negative action. We call this type of positive and negative energy "absolute positive energy" and "absolute negative energy," respectively. In contrast, we call the type of positive energy and negative energy recognized in current physics "relative positive energy" and "relative negative energy."

Absolute positive energy carries positive information. Absolute positive energy takes positive action. It brings connection, order, and certainty. It enhances our soul power, heart power, and mind power. It yields health, rejuvenation, longevity, joy, success, wisdom, harmony, and power.

Absolute negative energy carries negative information. Absolute negative energy generates negative action. It creates disconnection, disorder, and uncertainty. It decreases our soul power, heart power, and mind power. It causes sickness, decay, aging, disasters, pain, suffering, death, disempowerment, difficulties, challenges, and more.

Money is a kind of energy. If we obtain money through offering good services that enhance positive information, then the money will carry absolute positive energy. It will bring us joy, love, prosperity, wisdom, power, and more. If we obtain money through negative actions such as cheating, lying, stealing, and more, thc money will carry absolute negative energy. It will bring us sickness, aging, death, pain, sadness, suffering, broken relationships, financial challenges, and more. Paying attention to how we make money is crucial for bringing joy and success in every aspect of our lives.

Be aware of the kind of energy you are accumulating and using. This is critical for bringing success to every aspect of your life. At this moment, we invite you to pause. Go within to examine your life. What kind of energy are you using? Is it absolute positive energy or absolute negative energy? The kind of energy you are using determines the kind of life you have. When you apply absolute positive energy, you will have a life full of love, joy, success, wisdom, and beauty. If you apply absolute negative energy, you will endure sickness, disasters, pain, suffering, difficulties, challenges, decay, and even death. We wish you will always apply absolute positive energy and have success in every aspect of your life.

In summary:

- Relative positive energy brings more mobility and greater freedom.

- Relative negative energy brings restriction and makes one stuck.

- Absolute positive energy generates positive action, which brings connection and order. It enhances positive information.

- Absolute negative energy generates negative action, which brings disconnection and disorder. It enhances negative information.

- To have a healthy, happy, and successful life, it is important to apply absolute positive energy.

Positive Matter and Negative Matter

Physicists have discovered matter and antimatter. Particle physics studies the fundamental building blocks of matter in the universe, which have been found to be the so-called elementary particles, including electrons, photons, quarks, and more. Antimatter is matter that is made of antiparticles.

The theoretical physicist Paul Dirac predicted the existence of antiparticles from his mathematical formula, the Dirac equation. Four years later, the positron, the antimatter opposite of the electron, was observed in a laboratory. An antiparticle has the same qualities as the corresponding particle, except that it has the opposite charge. When a particle and its antiparticle (for example, an electron and a positron) meet, they will annihilate each other and become pure light. This pure light has no duality. In other words, antilight is the same as light.

In addition to matter and antimatter, Tao Science suggests the existence of positive matter and negative matter. Positive matter carries positive information. It is a positive transformer. It will bring more order and connection. It will enhance and uplift our souls, hearts, and minds. It will empower us and bring joy, love, success, wisdom, health, longevity, abundance, spiritual growth, and more to our lives. A book, music, movie, and food are positive matter if they carry positive information and bring more connection and order to our lives and the world. A healthy body is positive matter.

Negative matter is matter that carries negative information. It is a negative transformer. It will bring disorder and disconnection. It will decrease our soul, heart, and mind power. It will bring sickness, decay, aging, death, disasters, challenges, and difficulties. A book, music, movie, and food are negative matter if they carry negative information and bring disconnection and disorder to our lives and the world. If part of our body doesn't function well, it can be negative matter.

A house is positive matter if it carries positive information. For example, a house may have brought love and joy to many people. Many people cherish wonderful memories of this house. The house has offered much good service. This house carries positive information. Therefore, this house is positive matter. This house will then bring a lot of joy, flourishing, health, longevity, and other good fortune to the people who live in it now.

A house is negative matter if it carries negative information. For example, harmful actions may have taken place in a house in the past. In this case, this house carries negative information. It is negative matter. It may bring unhappiness, broken relationships, sickness, and misfortune to the people who live there now.

Matter is the transformer. When negative matter brings sickness, difficulties, challenges, and misfortune to us, it forces us to transform ourselves. In this way, matter has great power to help us grow, transform, and uplift our souls, hearts, minds, and every aspect of our lives. It is through matter, the transformer, that we transform our souls, hearts, and minds.

Having positive matter in our lives is essential for positive growth, positive transformation, and the upliftment of our souls, hearts, minds, and every aspect of our lives. Please take a moment now to evaluate the matter you have in your life. How much is positive matter? How much is negative matter? How can you transform the negative matter to positive matter?

The Power of Energy and Matter

Energy and matter have great value and significance for our souls, hearts, minds, and every aspect of our lives. We need to honor, respect, and cultivate the positive energy and matter in our lives.

Energy is action. Positive energy is positive power to act. Without positive energy, positive actions cannot happen. Taking positive actions is critical for a successful life. It is essential for enhancing our soul, heart, and mind power.

The one-sentence secret about using the power of energy is:

Take immediate action when receiving positive information.

By taking immediate action when we receive positive information, we are enhancing our souls, hearts, minds, and every aspect of our lives. Immediate action will give us more positive energy and matter immediately. If we don't take immediate action when we receive positive information, we will not be empowered to enhance our souls, hearts, minds, and every aspect of our lives. We will lose positive energy and matter. This is one of the important reasons why many people lack energy and even suffer from chronic fatigue. It is also why some people cannot flourish in their lives. We emphasize again: taking immediate action when receiving positive information is crucial for health and success in every aspect of our lives.

Matter is the physical manifestation of reality in our lives. Matter is the transformer of our souls, hearts, minds, and every aspect of our lives. The one-sentence secret about using the power of matter is:

We have the power to transform our souls, hearts, minds, and every aspect of our lives by using positive matter.

Many of us experience challenges, difficulties, sicknesses, and other blockages in our lives. It is crucial that we honor, respect, and be grateful for them. They are our teachers to teach us lessons. They appear to empower us. They come to help us heal, transform, and uplift our souls, hearts, minds, and every aspect of our lives.

The one-sentence secret about the power of energy and matter is:

By taking immediate action when we receive positive information and by using positive matter, we have the power to transform our souls, hearts, minds, and every aspect of our lives.

Let us be fully aware to bring and use positive energy and positive matter to take positive action and create positive transformation in every aspect of our lives.

How to Transform Negative Energy and Matter to Positive Energy and Matter

To transform negative energy and matter to positive energy and matter is to transform the negative information within the energy and matter into positive information. Whether energy or matter is positive or negative depends on your soul, heart, and mind. Do your soul, heart, and mind give, receive, and process positive information or negative information? By inputting, receiving, and processing positive information, we can transform negative energy and matter to positive energy and matter. In doing so, we can heal, transform, and uplift our souls, hearts, and minds and take our lives to a higher level.

Suppose you are facing a situation that brings you pain and suffering. You would like to transform this negative energy and matter into positive energy and matter so that you will have love and joy. How can you do this?

To turn a negative situation, which is negative energy and matter, into a positive situation, which is positive energy and matter, is to transform the negative information within the situation into positive information. We suggest the following four-step process:

Step 1. Transform negative soul power into positive soul power. Give positive information to this situation. Love melts all blockages and transforms every aspect of our lives. Love is the most powerful positive information. You can give love to this negative situation. Realize that this situation is occurring in your life to help you transform yourself and your life. It is your teacher to teach you something very valuable so that you can uplift your soul, heart, mind, and life. Feel the greatest love and blessings from this situation.

Step 2. Transform negative heart power into positive heart power. Transform frustration, sadness, irritation, selfishness, anger, anxiety, and other negative heart power to positive heart power. Feel the greatest love and gratitude for what is happening at this moment. Give your appreciation for this situation and everyone involved in it.

Step 3. Transform negative mind power into positive mind power. Ask what you can learn from this life situation. Transform negative mind-sets, negative beliefs, negative attitudes, negative habits, ego, attachment, and other negative mind power to positive mind power.

Step 4. Use what you learn from this situation to help others. The purpose of life is to increase our positive information. To increase our positive information is to enhance our connection with others. It is to empower us to serve better. To serve is to make others happier and healthier.

By completing the above four steps, you can transform any negative energy, matter, and situation to positive energy, matter, and situation. You could quickly heal, transform, and uplift every aspect of your life.

Chanting sacred mantras is a good way to transform negative energy and matter into positive energy and matter because it sends positive information and vibrations to us. To transform our negative energy and matter into positive energy and matter effectively, the secret is to chant with all of your soul, heart, and mind nonstop. If you can chant nonstop for a significant length of time, you can transform vast amounts of negative energy and matter to positive energy and matter.

One of the goals of Tao Science is to develop powerful techniques, practices, and technologies to help people transform negative energy and matter to positive energy and matter. In the next three chapters (seven, eight, and nine), we introduce Tao downloads, Tao transmissions, Tao Calligraphy, and more. These tools and techniques can help heal, transform, and uplift every aspect of your life quickly.

We encourage you to take some time now and regularly to examine every aspect of your life. Please check your thoughts, emotions, feelings, speech, writing, tasting, hearing, smelling, and other actions. Please check your books, music, home, relationships, career, finances, and more. You have the power to transform the negative energy and matter to positive energy and matter by inputting positive information. When you do so, you can heal, transform, and uplift every aspect of your life in a profound way.

How to Calculate Soul, Heart, and Mind

Matter has the power to transform our souls, hearts, and minds. In fact, from the composition of matter, we can calculate soul, heart, and mind mathematically. How? We are honored to explain this aspect of Tao Science.

Physics teaches us how to measure and calculate energy and matter. If we know how to calculate something, it means that we can grasp how to transform and fully utilize it. Because physics is able to measure and calculate energy and matter, it has great ability to transform, transfer, obtain, and utilize energy and matter. The results include all kinds of wonderful inventions and great feats, such as space travel, the internet, television, electric lights, lasers, and much more.

In Tao Science, we can calculate our soul, heart, and mind. The calculation of soul, heart, and mind will help us gain deeper insight and greater capability to develop and utilize our shen power, the power of our soul, heart, and mind. This calculation will also help us reach a better understanding of the highest soul, heart, mind, energy, and matter—what they are and how we can achieve them.

Using the mathematical tools developed in quantum physics, we can calculate our soul, heart, mind, energy, and matter. Let us now explain how to perform these calculations of soul, heart, and mind.

In quantum physics, everyone and everything is expressed mathematically by wave functions. A wave function describes the kinds and quantities of the vibrations in one's vibrational field. Quantum physics provides ways to calculate the wave function for everyone and everything. From the wave function, we can derive the behavior of everyone and everything.

For example, in quantum physics, we can calculate the wave function for the hydrogen atom. The hydrogen atom is the simplest and smallest atom. It is comprised of one electron orbiting around one proton. The hydrogen atom is one of the few systems whose wave function can be calculated exactly in quantum physics.

The mathematical expression of a hydrogen atom's wave function shows us that the electron within the atom circulates around the proton in certain discrete orbits. In other words, in the hydrogen atom's stable state, its electron does not move around the proton in random orbits. It is very similar to our solar system. The planets rotate around the sun in specific orbits. They do not rotate in random orbits. However, unlike the planets, the electron of a hydrogen atom can jump from one permissible orbit to another. In making this transition, the hydrogen atom either absorbs or emits light. If the electron jumps from a higher energy state to a lower energy state, a light wave will be emitted. For the electron to jump from a lower energy state to a higher energy state, the hydrogen atom must absorb a light wave that has energy exactly equal to the difference between the two energy states.

From the wave function of a hydrogen atom, we can derive what frequencies of light a hydrogen atom can emit and absorb. We can determine how the hydrogen atom will respond and behave if we shine light on it, heat it, or do anything to it. We can check our calculations using experimental measurements. In this way, we can learn everything we want to know about the hydrogen atom. We can reach a complete scientific understanding of the hydrogen atom.

How can we can calculate the soul, heart, and mind of a hydrogen atom? Soul is the information content of the atom. Information is the possible states of the atom. Since the wave function of a hydrogen atom tells us its possible states, we can determine the information content of the hydrogen atom from its wave function. Therefore, the wave function basically describes the soul of the atom.

To calculate the heart of a hydrogen atom, we need to know the types and quantities of vibrations the atom can receive. According to quantum physics, a hydrogen atom receives vibrations through a phenomenon called resonance. From its wave function, we can determine the kinds of vibrations a hydrogen atom can resonate with. In this way, we can know what kind of heart the hydrogen atom has. The heart of a hydrogen atom can be measured by a spectroscope. A spectroscope is an instrument that measures the kinds of light a hydrogen atom can receive or emit. The spectrum of light absorbed or emitted by a hydrogen atom represents the heart of the atom.

Next, how can we measure a hydrogen atom's mind? We can calculate or measure the atom's behavior or response after receiving a vibration. Since we know the wave function of a hydrogen atom, this calculation can be performed. The behavior or response is the atom's mind, which is the atom's consciousness.

The calculations of soul, heart, and mind outlined in the above paragraphs can be applied to any system. In principle, everyone and everything's soul, heart, and mind can be calculated if we know its wave function. At this time, computing the wave function of a complicated system is beyond human reach. However, with the development of quantum computing, the wave functions of more and more systems may become calculable.

The wave function can also be measured using detectors. From the calculation and/or measurement of more and more wave functions, the calculation of the soul, heart, mind, energy, and matter of more and more things can become possible in the future.

The measure of our soul is entropy and netrophy. Entropy measures how much negative information is within everyone and everything. Negative information measures the disorder and disconnection within our soul. Netrophy measures the positive information. Positive information measures the order within our soul and the correlation and connection our soul has with others. Our positive information or netrophy determines the power of our soul.

The measure of our heart is the amount of resonance our vibrational field can have with others. The more resonance we can feel with others, the more powerful our heart is. Various tools and detection methods, such as infrared and ultraviolet spectrometry, gamma ray detectors, magnetic resonance imaging (MRI), and electron microscopy, essentially measure the heart.

The measure of our mind is the quantity of information we can process per second in our vibrational field. The more information we can process per second, the more powerful our mind is.

In the next chapter, we will use the methods outlined here to discuss the highest form of soul, heart, mind, energy, and matter and how to reach this highest state. We first conclude this chapter by using what we have learned so far to address some controversies around quantum physics.

Resolution of Controversies around Quantum Physics

The basis of quantum physics is still not well understood by many physicists. This is mostly because it challenges the foundations and cornerstones of physics and natural science in some core issues: objectivity, predictability, and cause-effect of natural phenomena.

Quantum physicists study how to calculate the wave function of everyone and everything and derive the properties of anything from its wave function. Quantum physics uses wave functions to describe everyone and everything, while classical physics uses equations of motion to describe everyone and everything. This is the major difference between quantum physics and classical physics.

Equations of motion are deterministic. From an equation of motion, you can determine where an object was in the past and where it will be in the future. The nature of a wave function is probabilistic. A wave function can only tell us the possibilities inherent in the object. Because of its non-certain and non-deterministic nature, many physicists, including Albert Einstein, regard quantum physics as an unsatisfactory description of nature. Einstein considered the probabilistic nature of quantum physics to be an act of gambling. He could not believe that it represented the true and essential qualities of nature and God. Einstein died not believing in quantum physics because he could not believe that "God plays dice."

What Einstein and many scientists have not realized is that the probabilistic nature of quantum physics is not due to any imperfection of "God," nature, or quantum physics itself, but rather it is because of the profound truth and law of Tao Science about what everyone and everything is made of, the Law of Shen Qi Jing.

The Law of Shen Qi Jing, which is the Law of Information Energy Matter in scientific terms, tells us that everyone and everything is made of information, energy, and matter. Information describes the possibilities within everyone and everything. Information is intrinsically probabilistic. Therefore, the probabilistic nature of quantum physics and the wave function should no longer be surprising. A wave function tells us the possible states of a system, as well as the energy and matter carried by each state. This means that the wave function describes the information, energy, and matter within everyone and everything. This indicates that quantum physics follows the Law of Shen Qi Jing.

The subjective nature of quantum phenomena drastically conflicts with one of the cornerstones of scientific research and of natural science itself, which is objectivity. It is generally accepted in natural science that natural phenomena are objective. Their existence and occurrence do not depend on the observer. What happens does not depend on who is watching or how it is watched. The observation does not affect the event. In quantum physics, however, phenomena are subjective. They depend on the observer and on the action of the observer.

From our definitions of soul, heart, and mind in Tao Science, we can easily see that our heart and mind determine which potentialities or possibilities are manifested from our soul. That is why the reality we observe is subjective. The subjective nature of quantum phenomena can be easily comprehended. It confirms and illustrates scientifically what many spiritual teachers, such as Buddha, Jesus, and more, have been teaching us. All comes from our soul, heart, and mind.

The seeming violation of the cause-effect principle by the phenomenon of quantum entanglement can also be deduced from the Law of Shen Qi Jing. All examples of quantum entanglement are due to the fact that information, energy, and matter are the basic constituents of our existence. Space and time is a way to measure and organize our world. When we understand this, the existence of non-local phenomena is no longer surprising. The full derivation of quantum entanglement phenomena is related to another important law in Tao Science, the Law of Tao Yin Yang Creation. We will present the derivation in chapter eleven. As a corollary, this will show that the phenomenon of quantum entanglement does not violate cause-effect. Rather, it is actually the *result* of cause-effect.

In summary, the Law of Shen Qi Jing and the definitions of soul, heart, and mind give us a simple metaphysical understanding of quantum physics. In other words, Tao Science provides a way to understand quantum physics in terms everyone can understand. Quantum physics, in turn, provides a scientific proof and explanation of the Law of Shen Qi Jing as well as other spiritual wisdom and phenomena.

7

Grand Unified Field

A FIELD is something that extends over space and time. On Mother Earth, we can be in a field of grass, a field of flowers, a field of wheat, or a field of mountains. A river is a field. An ocean is a field. There are countless fields on Mother Earth.

In 1920, Einstein realized that a field, specifically the gravitational field, is a more accurate way to describe gravity than Newton's gravitational force. Quantum physics extends the concept of field to all things. It describes everyone and everything as a vibrational field made of various vibrations. A vibration is a wave. It is a periodic oscillation in space and time. It is by nature a field, as it extends over space and time.

The branch of physics called electromagnetism has shown that the electric force and the magnetic force are also the results of fields, respectively, the electric field and the magnetic field. Einstein's theory of special relativity shows that in four-dimensional spacetime, the electric and magnetic forces become one, and result from the electromagnetic field. In this way, the electric force and the magnetic force are unified in four-dimensional spacetime.

The unification of the electric force and the magnetic force supports Einstein's and much of humanity's deep belief that everything in nature comes from one source. Firmly convinced of this truth, Einstein spent much of his later years trying to find the mathematical formula for the unified field that includes both the gravitational field and the electromagnetic field. This started the continuing pursuit of the unified field and the unified theory.

With the development of quantum physics and the detection of the so-called weak force, strong force, and many elementary particles, the search for the unified field has been upgraded to the search for the grand unified field and the grand unified theory. Currently, string theory is able to unify all forces and all matter. However, it leaves some significant questions unanswered.

The search for the grand unified field is a deep quest to discover the source and origin of everyone and everything. It is the attempt to find the Creator and the Source scientifically. It is to use one mathematical formula to describe the Creator, Tao, and the Source. It is the effort to find the way to reach the Creator and the Source in a scientific way.

Is this possible?

Let us consider what insights spiritual wisdom can give us.

Calculation of the Grand Unified Field

Tao wisdom tells us that Tao, the emptiness, is the Creator and Source of everyone and everything. The grand unified field is the vibrational field of Tao.

In classical physics, emptiness is nothingness. There is literally nothing in the emptiness.

In quantum physics, emptiness is no longer nothingness. Vibrations can emerge from as well as go back into the emptiness. This phenomenon is called quantum fluctuation. Since there are no blockages in the emptiness, all sorts of vibrations can appear in the emptiness through quantum fluctuation.

A wave function describes the possible vibrations and states of something. The wave function can tell us the information, energy, and matter inside something. The renowned quantum physicist, Richard Feynman, discovered that to calculate the wave function is to sum up all possible states. If you attempt to calculate the wave function of emptiness, you will find that emptiness contains infinite possibilities and infinite vibrations. The infinite is something so huge or so small it cannot be counted or measured. It is bigger than biggest. It is smaller than smallest. Within the emptiness, there are countless information, energy, and matter.

Since physics studies only phenomena and matter in the observable universe, physics deals only with the measurable, which is inherently finite. Therefore, the infinite is one of the most challenging problems physicists have ever run into. In fact, encountering infinity is quite a scary experience for some physicists. At first, many physicists chose to ignore the infinite encountered in the emptiness. After all, nothing in this world is emptiness. However, they cannot shy away from the infinite and the emptiness for very long. Soon, they would find in quantum physics that matter interacts with the emptiness. This interaction can also create the infinite.

Nobel Laureate Paul Dirac is regarded as one of the most significant physicists of the twentieth century due to his mathematical brilliance. In the early 1980s, Paul Dirac told top string theorist Edward Witten at Princeton University that the most important challenge in physics was "to get rid of infinity."

For many decades, quantum physicists have been baffled about how to understand and deal with the infinite. Eventually, they developed a method called the renormalization procedure to deal with the infinity problem. We will not go into any details about renormalization. We simply want to summarize what we can learn from quantum physics about Tao, the grand unified field:

- Emptiness has boundless information, energy, and matter.
- Emptiness has infinite soul, heart, mind, energy, and matter.
- Emptiness is within everyone and everything.
- Emptiness interacts with and responds to everyone and everything.

Therefore, the grand unified field is indescribable. No numbers and no words can express this grand unified field.

The grand unified field has its soul, heart, mind, energy, and matter.

The soul of the grand unified field is boundless. It contains all possibilities and all information. It is quantum entangled with everyone and everything. It is connected with everyone and everything. It is within everyone and everything. It has the highest positive information.

The heart of the grand unified field is all-inclusive. It can receive and respond to any information from anyone and anything.

The mind of the grand unified field is limitless. It can process any information instantly.

The energy of the grand unified field is endless. It can never be exhausted.

The matter of the grand unified field is beyond countless. It is bigger than the biggest and smaller than the smallest.

The grand unified field is the ultimate life source that nourishes, rejuvenates, and energizes everyone and everything.

When we connect with this grand unified field, we can obtain any wealth, treasure, elixir, nectar, and much more. We can draw unlimited energy. We can gain any wisdom, knowledge, and secret, as well as any supernatural powers and abilities.

Connecting with the grand unified field is the next stage of human evolution for the highest information, unlimited energy, and countless matter.

At this moment, most of humanity remains at the level of consciousness that resources are limited. Therefore, one must struggle, compete, and fight with others for these resources. Most of humanity has not realized that unlimited information, energy, and matter are available to everyone right here and right now. We do not need to compete with anyone or fight with any group to obtain them. All we need to do is connect with the boundless grand unified field. All we would ever need, all we want, all we could ever imagine, all we cannot imagine, and all we have ever dreamed of can come to us. The purpose of this chapter is to share with you the wisdom, knowledge, and practices to empower you to achieve this.

In the future, technology will be developed and available for humanity and all beings to obtain unlimited energy, matter, and information from the grand unified field. We will never again need to worry about the energy crisis, financial challenges, and other shortages and sufferings. This type of technology will be manifested as soon as our souls, hearts, and minds are ready to bring it to reality.

How can we reach the grand unified field? Ancient wisdom provides important insights.

Spiritual Wisdom and the Grand Unified Field

Millions of people in history have searched for the truth about how our universe was formed, how it has developed, and how it will end. Many scientists have devoted their lives to search for this truth.

Ancient sacred spiritual wisdom has revealed this truth. In the classic text, *Dao De Jing*, Lao Zi states:

> *Tao Sheng Yi*
> *Yi Sheng Er*
> *Er Sheng San*
> *San Sheng Wan Wu*
>
> *Wan Wu Gui San*
> *San Gui Er*
> *Er Gui Yi*
> *Yi Gui Tao*

These two four-line stanzas describe Tao Normal Creation and Tao Reverse Creation, respectively.

Grand Unified Field, Tao Normal Creation, Tao Reverse Creation

Let's look at each line, starting with the four lines of Tao Normal Creation.

Tao Sheng Yi
Yi Sheng Er
Er Sheng San
San Sheng Wan Wu

Tao Sheng Yi

Tao is the Source, which is the Creator. Sheng means *creates*. Yi means *Oneness*. Tao Sheng Yi means *Tao creates Oneness*. Oneness is a field. This field is named the Hun Dun condition. Hun Dun means *blurred*. Tao creates the Hun Dun Oneness condition. In fact, Tao is the Hun Dun Oneness condition, and the Hun Dun Oneness condition is Tao. Within the Hun Dun Oneness condition there are two kinds of qi: qing qi (*light, pure energy*) and zhuo qi (*heavy, disturbed energy*). In the Hun Dun Oneness condition, qing qi and zhuo qi are blended and cannot be distinguished. They wait for eons until Tao decides it is time for qi transformation.

In Tao Science, the Hun Dun Oneness condition is the grand unified field.

Yi Sheng Er

Er means *two*. Yi Sheng Er means *Oneness creates Two*. Yi is the Hun Dun Oneness field, which is the grand unified field. When qi transformation occurs, the light, pure energy rises to form Heaven. The heavy, disturbed energy falls to form Earth. Heaven and Earth are two. Heaven is yang. Earth is yin.

In Tao Science, Heaven is a quantum field. Earth is also a quantum field.

Er Sheng San

San means *three.* Er Sheng San means *Two creates Three.* Two are Heaven and Earth. Three are the Hun Dun Oneness condition plus Heaven and Earth.

In Tao Science, this Three is also a quantum field.

San Sheng Wan Wu

San means *three.* Sheng means *creates.* Wan means *ten thousand.* In Chinese, ten thousand represents infinity. Wu means *things.* San Sheng Wan Wu means *Three creates countless planets, stars, galaxies, and universes.* Mother Earth is only one planet.

In Tao Science, each planet is a quantum field. Each star, each galaxy, each universe, and each human being is a quantum field. Everyone and everything is a quantum field. The Hun Dun Oneness condition is the grand unified field.

To explain grand unification further, we must know another ancient sacred wisdom that is now brought to the twenty-first century to explain other universal truths and help unify science and spirituality. This is the Law of Shen Qi Jing that we explained in chapter three: everyone and everything is made of shen qi jing. Recall that jing means *matter.* Qi means *energy.* Shen includes *soul, heart,* and *mind.*

In Tao Science, jing or matter is our physical existence. Qi is energy, which has the function to move matter. Shen is information, which includes three aspects: content of information (soul), receiver of information (heart), and processor of information (mind).

Tao Sheng Yi, Yi Sheng Er, Er Sheng San, San Sheng Wan Wu is Tao Normal Creation.

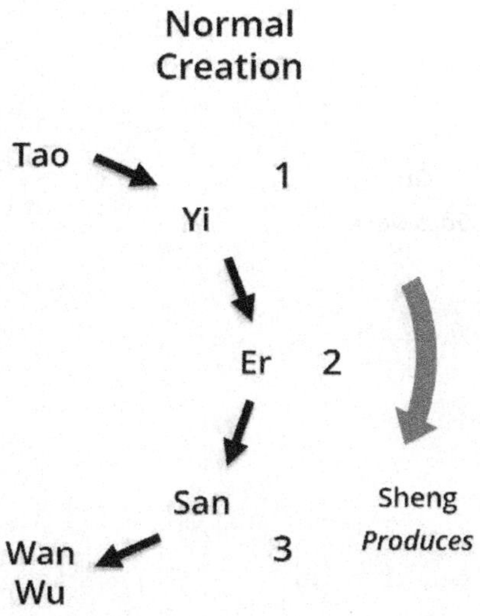

Figure 3. Tao Normal Creation

See figure 3. Tao creates One. One creates Two. Two creates Three. Three creates everyone and everything. This process is named Tao Normal Creation.

At the first step in Tao Normal Creation, Tao "creates" the Hun Dun Oneness condition. In fact, they are one and the same. Tao and the Hun Dun Oneness condition belong to the Wu World, which is the realm of emptiness and nothingness. In ancient wisdom, Wu creates You. The You World is the realm of existence. It includes Two, Three, and wan wu. Wan wu contains countless planets, stars, galaxies, and universes, as well as human beings, animals, plants, minerals, cells, molecules, atoms, electrons, quarks, and everyone and everything.

Tao Normal Creation explains how the universe (the You World of wan wu, or everyone and everything) was created. How will the universe develop and end? Let's look at the four lines in Tao Reverse Creation. See figure 4 on the next page.

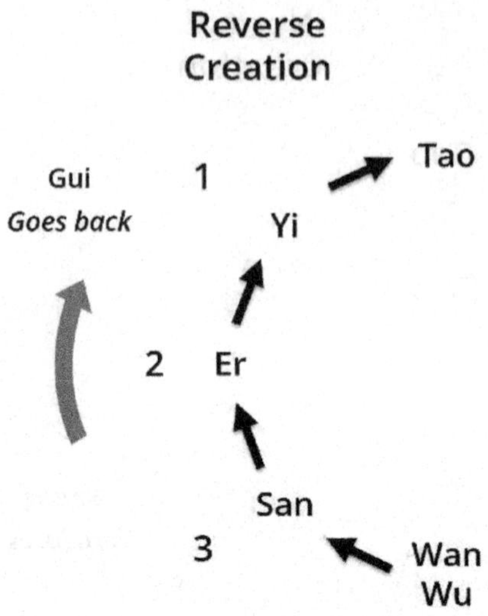

Figure 4. Tao Reverse Creation

Wan Wu Gui San
San Gui Er
Er Gui Yi
Yi Gui Tao

Wan Wu Gui San

Wan Wu means *ten thousand things*, which is a way to express all things. Gui means *go back*. San means *three*. Wan Wu Gui San means *Countless planets, stars, galaxies, universes, and human beings go back to Three.*

San Gui Er

San Gui Er means *Three goes back to Two.* Recall that this *Three* is the Hun Dun Oneness condition plus Heaven and Earth.

Er Gui Yi

Er Gui Yi means *Two goes back to One.* This *Two* is Heaven and Earth. This *One* is the Hun Dun Oneness condition.

Yi Gui Tao

Yi Gui Tao means *One goes back to Tao.*

In Tao Science, we can explain the wisdom scientifically. Wan wu is a quantum field. It is made of jing qi shen. "Three" is a quantum field. It is made of jing qi shen. "Two" is a quantum field. It is made of jing qi shen. "One" is the Hun Dun Oneness field, which is the grand unified field. Tao creates and is the Hun Dun Oneness field, which is also the grand unified field.

From Tao Normal Creation and Tao Reverse Creation, we can summarize and infer as follows:

- Tao creates One, which is the grand unified field.
- Heaven and Earth are two quantum fields.
- Wan wu, including countless planets, stars, galaxies, universes, and human beings, is countless quantum fields.
- Tao and One are emptiness and nothingness that contain countless positive information, energy, and matter. Heaven and Earth contain much less positive information, energy, and matter than Tao and One. Countless planets, stars, galaxies, universes, and human beings carry less positive information, energy, and matter than Heaven and Earth.
- Wan Wu Gui San, San Gui Er, Er Gui Yi, Yi Gui Tao is Tao Reverse Creation. It is exactly the reverse of Tao Normal Creation.

- Within Tao Normal Creation, Tao and One carry the highest countless, purest, and most positive information, energy, and matter. Heaven and Earth carry less. Wan wu carries even less.

- Within Tao Reverse Creation, wan wu will purify and transform their information, energy, and matter to reach the purity of the information, energy, and matter of Heaven and Earth. Then, the information, energy, and matter of Heaven and Earth will purify and transform further, to reach and return back to the highest purity of the information, energy, and matter of Tao and One.

- In Tao Science, we can state that Tao and One carry the most negative entropy and the most positive information. Within Tao and One, there is no negative information. Heaven, Earth, and wan wu have progressively less and less pure and positive information. Therefore, Heaven, Earth, and wan wu carry more and more entropy and negative information.

Tao Reverse Creation is the process of transforming negative information, energy, and matter to positive information, energy, and matter. Wan wu carries the most negative information, energy, and matter. Wan wu going back to Heaven and Earth is wan wu purifying and reducing negative information and entropy, and increasing positive information and negative entropy.

When Heaven and Earth go back to One and Tao, their information has become completely positive. A state of complete negative entropy is reached.

Grand Unification Formula

More than five thousand years ago, ancient Taoist masters already knew that there is a Source that existed before Heaven and Earth. This Source created Heaven, Earth, everyone, and everything. They found that the secret to connect with this Source is to bring shen, qi, and jing together as one. This profound insight has helped many people reach high-level physical and spiritual achievement. This deep wisdom is also the gateway to find and reach the grand unified field.

In 2013, while working together, Dr. Rulin shared with me (Master Sha) her inspiration of obtaining the grand unified theory, the one formula that includes everyone and everything. I closed my eyes for a minute. Then, I wrote on a piece of paper the grand unification formula:

S + E + M = 1

S denotes shen, which includes soul, heart, and mind—respectively, the content, receiver, and processor of information. E is energy, which has the function to move matter. M is matter, which is the physical existence of everyone and everything.

The grand unified field or Grand Unification Theory (GUT) tells us that everyone and everything is made of shen qi jing. Tao and One are the ultimate Creator. Heaven and Earth (yang and yin) are secondary creators. Three is the next layer of creator. Wan wu is yet another layer of creator. We are all creators, but we are different layers of creators.

Tao Normal Creation explains how everyone and everything is created. Tao Reverse Creation explains how everyone and everything develops and ends at their ultimate destiny.

Human beings and wan wu have challenges and difficulties because their shen qi jing are not aligned as one. In other words, their information, energy, and matter are not aligned as one. They are separated, disconnected, and disordered.

Every aspect of life follows Tao Normal Creation and Tao Reverse Creation because that is the highest philosophy, highest theory, highest practice, highest science, and highest principles and laws.

Millions of people are sick.

Millions of people are dying in hospitals.

Millions of people have relationship challenges.

Millions of people have financial challenges.

Millions of people have success challenges.

Millions of people have all kinds of challenges.

Why? See the grand unification formula, S + E + M = 1. Everyone and everything, and every aspect of life, are made of shen qi jing. All kinds of challenges and difficulties are due to shen qi jing not being joined as one.

Shen could have blockages. Qi could have blockages. Jing could have blockages. In Tao Science, information, energy, and matter all could have blockages. To transform all of these blockages is to align shen qi shen as one. Then, we can truly go back to One and Tao.

Why is this formula the grand unification formula? It can explain how countless planets, stars, galaxies, universes, and human beings are formed, develop, and end. It can also explain how to transform health, relationship, finances, success, science, business, every occupation, rejuvenation, longevity, immortality, and every aspect of life. Therefore, in one sentence:

S + E + M = 1 is the scientific equation of Grand Unification Theory and practice.

S + E + M = 1 holds the solution to transform everyone and everything, from the biggest universe to the smallest quark, in every aspect of life. The scientific insights and practical benefits are immeasurable. To apply this grand unification formula is to offer revolutionary service to transform humanity, Mother Earth, and countless planets, stars, galaxies, and universes.

Apply S + E + M = 1 to Transform All Life

We have taught the Four Power Techniques to transform all life. For thousands of years, spiritual and energy practitioners have used three of these sacred techniques. Now, we introduce six techniques to transform all life.

1. **Body Power** (known since ancient times as Shen Mi, *body secrets*)

   Body Power is to use body and hand positions for healing and transformation. In one sentence, where you put your hands is where you receive healing and transformation.

2. **Sound Power** (known since ancient times as Kou Mi, *mouth secrets*)

   Sound Power is to chant special sounds or messages, such as ancient and modern sacred mantras, that carry positive information, energy, and matter, which can transform negative information, energy, and matter.

3. **Mind Power** (known since ancient times as Yi Mi, *thinking secrets*)

   Mind is consciousness. There are many kinds of consciousness, including superficial consciousness, deep consciousness, subconsciousness, superconsciousness, and much more. Mind Power is the power of consciousness.

4. **Soul Power**

   Soul Power is the power of information. Spiritual beings speak about soul or spirit. Quantum science speaks about information or message. They are different words for the same thing.

   In Tao Science, soul power is divided into positive soul power and negative soul power. Soul is information, which can be divided into positive information and negative information. Positive information is positive karma, which is measured by negative entropy. Negative information is negative karma, which is measured by entropy. Soul Power is to apply positive information to transform all kinds of negative information.

   In Tao Science, we use the grand unification formula, S + E + M = 1, to transform every aspect of life. S + E + M = 1 carries Tao and Hun Dun Oneness field information, which is the purest, most positive, limitless information. It has the highest negative entropy. Later in this chapter, we will lead you to apply S + E + M = 1 to transform all life.

5. **Breathing Power**

Many teachers have taught many breathing power techniques. We emphasize one sacred breathing power technique to transform all life. When you meditate, inhale and visualize a light channel flowing from the navel to the Ming Men acupuncture point, which is the acupuncture point located on the Du meridian (Governing Vessel) on the back, directly behind the navel. Ming means *life*. Men means *gate*. Therefore, the Ming Men point is the *life gate*. In traditional Chinese medicine, the Ming Men point is a hub for the Five Elements (Wood, Fire, Earth, Metal, Water) and the four extremities.

The Wood element includes the liver, gallbladder, eyes, tendons, and anger in the emotional body.

The Fire element includes the heart, small intestine, tongue, blood vessels, and depression and anxiety in the emotional body.

The Earth element includes the spleen, stomach, mouth, lips, gums, teeth, muscles, and worry in the emotional body.

The Metal element includes the lungs, large intestine, nose, skin, and grief or sadness in the emotional body.

The Water element includes the kidneys, urinary bladder, ears, bones, and fear in the emotional body.

The most sacred wisdom is that Tao creates everyone and everything. Where is Tao within the body? The Ming Men point is the Tao point of the body. To inhale while visualizing light flowing from the navel to the

Ming Men point is to connect with Tao. When you exhale, you can visualize light flowing back from the Ming Men point to the navel. You are connecting with Tao Normal Creation and Tao Reverse Creation, which is to connect with Tao, One, Two, Three, and wan wu.

6. **Tracing and Writing Power—Tao Calligraphy**

   Tao Calligraphy is a unique form of Chinese calligraphy created by Master Sha. It is Oneness writing. A single Chinese character is traditionally written with one stroke up to more than twenty strokes. Oneness writing is to connect every stroke in a character, or even several characters in an entire phrase, as one. Every stroke is made of shen qi jing. To connect every stroke as one is S + E + M = 1.

   We will focus on Tao Calligraphy and Tracing and Writing Power in chapter nine.

Power and Significance of the Grand Unified Field

In Tao Science, we understand that everyone and everything are made of jing qi shen. A human being is made of jing qi shen. An animal is made of jing qi shen. An ocean, a mountain, a tree, a flower, a house, an organization, a city, a country—all are made of jing qi shen. Mother Earth is made of jing qi shen. Countless planets, stars, galaxies, and universes are all made of jing qi shen. In one sentence, everyone and everything is a quantum field of jing qi shen.

Additional important wisdom of Tao Science is that every field is made of information, energy, and matter. We would like to emphasize what we explained earlier: Heaven, Earth, countless planets, stars, galaxies, and universes, and human beings all have jing qi shen that are not aligned as one. Tao

Science would say that their information, energy, and matter are not aligned as one. Therefore, everyone and everything has limitations.

Compared to Heaven and Earth, the lifespan of a human being is extremely limited. Few human beings live more than one hundred years. Heaven and Mother Earth have lived for billions of years. Therefore, a human's life is limited. The lives of Heaven and Earth are unlimited.

Compared to Tao (Hun Dun Oneness field), the lives of Heaven and Earth are limited. Tao Oneness is unlimited. Tao is beyond immortality. Tao has no beginning and no ending.

Why do human beings, Earth, Heaven, and Tao have different lifespans? It is because they have different jing qi shen. In Tao Science, information, energy, and matter can have countless layers of purity, frequency, and vibration. Tao information, energy, and matter carry the highest possible purity, the most complete connectedness, and the ultimate absolute order. Tao has complete negative entropy. Heaven and Earth are in different layers, with less purity, connectedness, and order. Human beings are in different layers again, with even less purity, connectedness, and order.

Grand unification is the Tao Oneness field. The grand unification formula, S + E + M = 1, is to reach the Tao Oneness field. Heaven, Earth, human beings, everyone, and everything have not yet reached S + E + M = 1. The theory and practice of the grand unification formula are to apply the formula to reach the Tao Oneness field. Because of many impurities, it could take a long time to reach the Tao Oneness field.

We cannot emphasize enough that the Tao Oneness field carries the purest jing qi shen. The Tao Oneness field carries the purest matter, energy, and information. The Tao Oneness field carries complete negative entropy. Therefore, to transform anyone and anything completely, including countless planets, stars, galaxies, universes, and human beings, is to reach the Tao Oneness field. In one sentence:

The grand unification formula, S + E + M = 1, is the Tao Source Oneness treasure and Tao Science treasure to transform everyone and everything.

How do we use the grand unification formula for transformation? We will show you now.

Apply the Grand Unification Formula to Heal

Apply the Four Power Techniques and the grand unification formula, S + E + M = 1, to heal:

Body Power. Put one palm on your Ming Men acupuncture point (on your back directly behind the navel). Put your other hand on any part of the body that needs healing.

Soul Power. Say *hello* to inner souls:

Dear soul mind body of my Ming Men acupuncture point, my Tao Oneness point,
I love you.
You have the power to heal me.
Do a good job.
I am very grateful.

Say *hello* to outer souls:

The key to invoking the outer soul of S + E + M = 1 is forgiveness practice. Forgiveness practice transforms negative information. This negative information is shen qi jing blockages.

To transform this negative information is to self-clear negative karma, heart blockages, mind blockages, energy blockages, and matter blockages.

Here is how to use the sacred Tao Science grand unification formula, S + E + M = 1, to transform negative information, including self-clearing negative karma:

> *Dear Tao Science grand unification formula*
> *S + E + M = 1,*
> *I love you.*
> *You have the power to forgive my ancestors and me for our mistakes, which are the negative information we have created in all lifetimes.*
> *Please forgive us and transform our negative shen qi jing, which is negative information, energy, and matter, to positive shen qi jing, and also transform our entropy to negative entropy.*
> *I am extremely grateful.*

Sound Power. Chant repeatedly, silently or aloud:

> *S + E + M = 1*
> *S + E + M = 1*
> *S + E + M = 1*
> *S + E + M = 1 ...*

Chant for at least ten minutes per practice. You can practice several times a day. For chronic and life-threatening conditions, chant for a total of two hours or more each day. There is no time limit. The longer you practice, the better the results you could achieve. Thousands of people have received heart-touching and heart-moving healing results by practicing with S + E + M = 1.

Mind Power. As you chant $S + E + M = 1$, visualize golden light shining in the area where you requested healing.

Rejuvenation, Longevity, and Immortality

Millions of people worldwide and billions of people in history have sought and treasured rejuvenation and longevity. Going beyond longevity, reaching immortality has also been a dream of many.

Remember Tao Normal Creation:

> Tao creates One.
> One creates Two.
> Two creates Three.
> Three creates wan wu.

Where is a human being? Human beings are in the wan wu level. Why do people age and get sick, making longevity difficult and immortality impossible? In the wan wu level, we are far from Heaven and Earth. We are much farther from Tao. We carry too much negative information, energy, and matter in our souls, hearts, minds, and bodies. We lack the purity and positive information, energy, and matter that would purify and uplift our shen qi jing to the quality of Mother Earth's shen qi jing, Heaven's shen qi jing, and Tao shen qi jing. Therefore, we cannot live long lives with good health and youthful vigor, and immortality remains an unattainable dream.

Two is yin and yang, Earth and Heaven. In ancient wisdom, Heaven and Earth are our parents. We have physical parents. We also have spiritual parents. Your father and mother had intercourse, and your father's sperm and mother's egg joined to form a zygote. Nine months later, you were born. You may not have heard that Heaven and Earth also interact. They interacted millennia ago, even eons ago to create a soul. This soul is your soul, which enters your body the moment you take your first breath after emerging from your mother's womb. Heaven is our father. Earth is our mother. Without Heaven and Earth, there would be no human beings.

Lao Zi said, "Ren Fa Di. Di Fa Tian. Tian Fa Tao. Tao Fa Zi Ran." Ren means *human being*. Fa means *follow* or *transform*. Di means *Mother Earth*. Tian means *Heaven*. Tao is the Source. Zi Ran means the *natural world*. These four sentences carry wisdom beyond comprehension.

Ren Fa Di means that human beings must *follow the natural laws of Mother Earth*. For example, when it rains, we need an umbrella. In the winter, we must wear winter clothes. There is hidden wisdom within this sacred phrase.

As we have shared, everyone and everything are made of jing qi shen. A human being has human jing qi shen. Mother Earth has Earth jing qi shen. Human jing qi shen is quite different from Earth jing qi shen. From the perspective of Tao Science, information, energy, and matter can all be divided between positive and negative. The positive information, energy, and matter of Heaven and Earth are much greater than a human being's. Therefore, a human's age is very limited. Heaven and Mother Earth have lived much longer.

Ren Fa Di has profound sacred hidden wisdom. It speaks to Tao Reverse Creation. The first step in Tao Reverse Creation is to transform a human's jing qi shen to Mother Earth's jing qi shen. In Tao Science, it is to transform and purify information, energy, and matter from the human being level to the Mother Earth level. It is also to transform entropy to negative entropy from the human being level to the Mother Earth level. This is a scientific explanation of the process of rejuvenation, longevity, and immortality.

Ren Fa Di, Di Fa Tian is the process of rejuvenation and longevity. Di Fa Tian means *Mother Earth must follow Heaven's rules.* In Tao Science, Di Fa Tian is to purify and transform the jing qi shen of Mother Earth to the jing qi shen of Heaven. In other words, it is to transform information, energy, and matter from Mother Earth's level to Heaven's level.

Tian Fa Tao is the process to reach immortality. Tian Fa Tao means *Heaven must follow Tao's rules.* Tao is the ultimate Source and Creator. Tao is the universal principles and laws that Heaven, Mother Earth, human beings, and all souls must follow. Tao is the emptiness that carries all energy and matter. It is the purest jing qi shen, which is complete positive information, energy, and matter. It has the highest negative entropy. Tao has no beginning and no ending.

Tian Fa Tao is to transform the jing qi shen of Heaven to the jing qi shen of Tao Source. In Tao Science, we explain that this is to purify and transform the information, energy, and matter of Heaven to that of Tao Source.

To reach Tao is to reach immortality. Because human beings cannot see or know the immortals, very few people truly believe in the possibility of immortality. The immortals are great but humble servants. They will not and need not reveal their identity to others. They serve in quiet anonymity.

Tao Science explains why immortality is possible and tells us what is required to achieve it. As you can imagine, it is not easy to reach immortality. It requires purification of our jing qi shen beyond any imagination. There are layers and layers of purification. Within each of the steps—Ren Fa Di, Di Fa Tian, Tian Fa Tao—to reach Tao, there are countless layers of purification.

Lao Zi shared many profound secrets. After Tian Fa Tao, he revealed still another sacred phrase: Tao Fa Zi Ran. Tao is the Source. Fa means *method* or *way* or *law*. Zi Ran means *nature*. In ancient wisdom, Tao is also named "nature." Tao Fa Zi Ran can be translated as *follow nature's way*. To reach Tao, one must follow Tao. To follow Tao is to follow nature's way. Shun Tao Chang, Ni Tao Wang. *Follow Tao, flourish. Go against Tao, finish.*

Lao Zi is recognized and respected worldwide as a sage. His philosophy and his greatest work, *Dao De Jing*, are recognized in major universities worldwide. For centuries, philosophers, scientists, politicians, economists, and more have studied *Dao De Jing*. Lao Zi's words in *Dao De Jing* may be simple and brief, but they are not easy to understand. Millions of readers of *Dao De Jing* in history do not have a clear or deep understanding of what Lao Zi was trying to explain.

Tao Science can help explain Lao Zi's wisdom. Ren (*human being*) lives in the third dimension. Di (*Mother Earth*) is currently in the third dimension. Mother Earth could turn to the fourth dimension. Through spiritual communication, we have learned that Mother Earth will uplift its frequency to the fourth dimension around the year 2150.

Why have millions and billions of people in history had difficulties to understand Ren Fa Di, Di Fa Tian, Tian Fa Tao, Tao Fa Zi Ran? One reason is that Ren, Di, Tian, and Tao are in completely different dimensions. Ren, Di, and Tian have space and time. Tao has no space and no time. Heaven has countless dimensions. Heaven has countless layers. Tao is beyond Heaven. It is the ultimate Creator. Tao carries beyond infinite dimensions. In Tao Science, the process of Ren Fa Di, Di Fa Tian, Tian Fa Tao is to purify and transform information, energy, and matter further and further. Tao Fa Zi Ran is the immortal state.

With Ren Fa Di, Di Fa Tian, Tian Fa Tao, Tao Fa Zi Ran, Lao Zi gave us the four steps to reach immortality. This path is the path of Tao Reverse Creation. (See figure 4 on page 124.) Wan Wu Gui San, San Gui Er, Er Gui Yi, Yi Gui Tao is the same path to immortality, using different words.

Sacred Practice for Rejuvenation, Longevity, and Immortality

Theory and practice are two. They are a yin-yang pair. In history, billions of people have studied *Dao De Jing*. Most of them believe Ren Fa Di, Di Fa Tian, Tian Fa Tao, Tao Fa Zi Ran are theoretical statements. In Tao Science, we use these sacred phrases as practices. Theory and practice are one. Now, we release the sacred practice of these four phrases to empower you and humanity for rejuvenation, longevity, and immortality.

In one sentence, Tao Normal Creation is Tao *creates everyone and everything*; Tao Reverse Creation is *everyone and everything go back to Tao*. See figure 5.

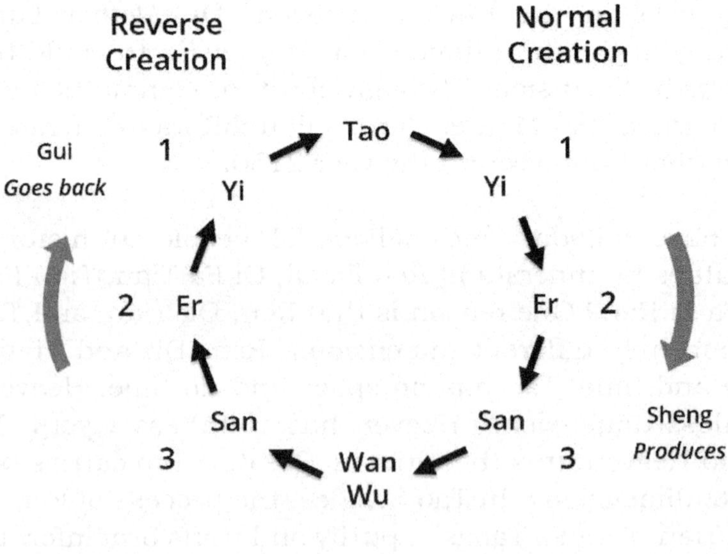

Figure 5. Tao Normal Creation and Tao Reverse Creation

Practice with us now for your journey of rejuvenation, longevity, and immortality.

Apply the Four Power Techniques.

Body Power. Grasp your left thumb with your right palm. The tip of your left thumb should touch the crease in your right palm below your right ring finger. This Body Power secret is called the Yin-Yang Palm. See figure 6.

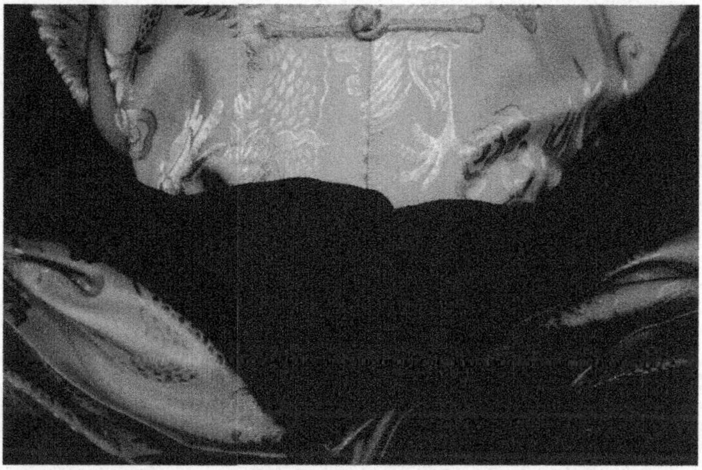

Figure 6. Yin-Yang Palm Body Power secret

Soul Power. Say *hello* to inner souls:

Dear my shen qi jing of every system, every organ,
every tissue, every cell, every DNA, and every
RNA in my body, from head to toe, skin to bone,

I love you, honor you, and appreciate you.

You have the power to heal, transform, rejuvenate,
prolong life, and move toward immortality.

Do a great job!

Thank you.

Say *hello* to outer souls:

Dear Tao, the Divine, Heaven, Earth,
Please forgive my ancestors and me for all the mistakes we have made in all lifetimes.
These mistakes include killing, harming, stealing, taking advantage of others, and more.
These mistakes are the negative information, energy, and matter that my ancestors and I carry.
They are the entropy.
They are the reasons for sickness and aging.
We sincerely ask for forgiveness.
We know in our hearts and souls that only to ask for forgiveness is not enough.
We have to serve unconditionally.
To serve is to make others happier and healthier.
Tao Science explains that to serve is literally to increase positive information, energy, and matter.
It is also to transform entropy to negative entropy.
This is how can we become younger.
This is how we can prolong our lives.
This is the possibility for us to reach immortality.

Sound Power. Chant repeatedly, silently or aloud:

Tao Sheng Yi, Yi Sheng Er, Er Sheng San, San Sheng Wan Wu (pronounced *dow shung ee, ee shung ur, ur shung sahn, sahn shung wahn woo*)

Wan Wu Gui San, San Gui Er, Er Gui Yi, Yi Gui Tao (pronounced *wahn woo gway sahn, sahn gway ur, ur gway ee, ee gway dow*)

Tao Sheng Yi, Yi Sheng Er, Er Sheng San, San Sheng Wan Wu
Wan Wu Gui San, San Gui Er, Er Gui Yi, Yi Gui Tao

Tao Sheng Yi, Yi Sheng Er, Er Sheng San, San Sheng Wan Wu
Wan Wu Gui San, San Gui Er, Er Gui Yi, Yi Gui Tao

Tao Sheng Yi, Yi Sheng Er, Er Sheng San, San Sheng Wan Wu
Wan Wu Gui San, San Gui Er, Er Gui Yi, Yi Gui Tao ...

Another mantra to chant for rejuvenation, longevity, and immortality is *Shen Qi Jing He Yi* (pronounced *shun chee jing huh ee)*. He means *join as*. Yi means *one*. Shen Qi Jing He Yi is Chinese for S + E + M = 1.

Chant now:

Shen Qi Jing He Yi, S + E + M = 1
Shen Qi Jing He Yi, S + E + M = 1
Shen Qi Jing He Yi, S + E + M = 1
Shen Qi Jing He Yi, S + E + M = 1 ...

We suggest that you alternate these two ways to chant in your practice. Chant Tao Normal Creation and Tao Reverse Creation one day, and chant *Shen Qi Jing He Yi, S + E + M = 1* the next day. We will introduce another powerful practice with Shen Qi Jing He Yi in chapter nine.

Chant for ten minutes or more every time you practice. In fact, for rejuvenation and longevity, chant at least one hour per day. Serious seekers of longevity and immortality should chant at least three hours per day. Add all of your daily practice time together to reach the target. There is no limitation. The longer you chant, the better the results you could receive.

To chant Tao Normal Creation and Tao Reverse Creation is to connect with Ren Di Tian Tao. Ren Di Tian Tao are all made of jing qi shen. Ren Di Tian Tao are all different fields.

To chant these sacred phrases is to purify and transform our negative information, energy, and matter to positive information, energy, and matter. It is also to transform entropy to negative entropy, from head to toe, skin to bone. Chant as much as you can.

8

Divine and Tao Spiritual Treasures

IN JULY 2003, I (Master Sha) was leading a Soul Study workshop near Toronto. The Divine came, as I could see clearly with my Third Eye. I explained to my students that the Divine had appeared, and asked them to wait a moment while I bowed down one hundred eight times to the Divine and waited for the Divine's message.

At age six, I was taught to bow down to my tai chi master. At age ten, I bowed down to my qi gong master. At age twelve, I bowed down to my kung fu master. As a Chinese, I learned the importance of this courtesy throughout my childhood.

I was honored to hear the Divine tell me, "Zhi Gang, I come today to choose you as my direct servant, vehicle, and channel."

I was deeply moved and replied to the Divine, "I am honored. What does it mean to be your direct servant, vehicle, and channel?"

The Divine explained. "When you offer healing and blessing to others, call me. I will come instantly to offer my healing and blessing to them."

I was even more deeply touched and could only say, "Thank you so much for choosing me as your direct servant."

The Divine continued, "I can offer my healing and blessing by transmitting my permanent healing and blessing treasures."

I asked with wonder, "How do you do this?"

The Divine answered, "Select a student and I will give you a demonstration."

I asked for a volunteer with serious health challenges. A man named Walter raised his hand. He stood up and explained that he had just been diagnosed with liver cancer, with a two-by-three-centimeter malignant tumor in his liver.

I then asked the Divine, "Please bless Walter. Please show me how you transmit your permanent treasures." Immediately, I saw the Divine send a beam of light from the Divine's heart to Walter's liver. The beam shot into his liver, where it turned into a golden light ball that instantly started spinning. Walter's entire liver shone with beautiful golden light.

The Divine then asked me, "Do you understand what software is?"

I was surprised by this question, but replied, "I do not understand much about computers. I just know that software is a computer program. I have heard about accounting software, office software, and graphic design software."

"Yes," the Divine said. "Software is a program. Because you asked me to, I transmitted and downloaded my Soul Software for Liver to Walter. It is one of my permanent healing and blessing treasures. You asked me. I did the job. This is what it means for you to be my chosen direct servant and channel."

I was astonished. Excited, inspired, and humbled, I said to the Divine, "I am so honored to be your direct servant. How blessed I am to be chosen." Almost speechless, I asked the Divine, "Why did you choose me?"

"I chose you," said the Divine, "because you have served humanity for more than one thousand lifetimes. You have been very committed to serving my mission through all your lifetimes. I am choosing you in this life to be my direct servant. You will transmit countless permanent healing and blessing treasures from me to humanity and all souls. This is the honor I give to you now."

I was moved to tears. I immediately bowed down to the Divine one hundred eight times again and made a silent vow:

> *Dear Divine,*
>
> *I cannot bow down to you enough for the honor you have given to me. No words can express my greatest gratitude. How blessed I am to be your direct servant to download your permanent healing and blessing treasures to humanity and all souls. Humanity and all souls will receive your huge blessings through my service as your direct servant. I give my total life to you and to humanity. I will accomplish your task. I will be a pure servant to humanity and all souls.*

I bowed again, and then I asked the Divine, "How should Walter use his Soul Software?"

"Walter must spend time to practice with my Soul Software," said the Divine. "Tell him that simply to receive my Soul Software does not mean he will recover. He must practice with his treasure every day to restore his health, step by step."

I asked, "How should he practice?"

The Divine gave me this guidance: "Tell Walter to chant repeatedly: *Divine Liver Soul Software heals me. Divine Liver Soul Software heals me. Divine Liver Soul Software heals me. Divine Liver Soul Software heals me.*"

I asked, "For how long should Walter chant?"

The Divine answered, "At least two hours a day. The longer he practices, the better. If Walter does this, he could recover in three to six months."

I shared this information with Walter, who was excited and deeply moved. Walter said, "I will practice two hours or more every day."

Finally, I asked the Divine, "How does the Soul Software work?"

The Divine replied, "My Soul Software is a golden healing ball that rotates and clears shen qi jing blockages from Walter's liver."

I again bowed to the Divine one hundred eight times. Then, I stood up and offered three Soul Softwares to every student in the workshop as divine gifts. Upon seeing this, the Divine smiled and left.

Walter immediately began to practice as directed for at least two hours every day. Two-and-a-half months later, a CT scan and MRI showed that his liver cancer had completely disappeared. At the end of 2006, I met Walter again at a book signing in Toronto for my book, *Soul Mind Body Medicine.*[8] In

---

[8] *Soul Mind Body Medicine: A Complete Soul Healing System for Optimum Health and Vitality.* Novato: New World Library, 2006.

May 2008, Walter attended one of my events at the Unity Church of Truth in Toronto. On both occasions, Walter told me that there was still no sign of cancer in his liver. For five years, the Divine Soul Software downloaded to his liver healed his liver cancer. He was very grateful to the Divine.

This is how I started to offer Divine treasures. There are countless Divine treasures, including Divine Shen Qi Jing light balls for bodily systems, organs, and cells. There are also many varied Divine treasures for transforming health, relationships, and finances, for increasing intelligence and opening spiritual channels, and for bringing enlightenment. For example, I have transmitted divine soul light acupuncture needles, divine soul light herbs, and much more.

Divine Heart Shen Qi Jing He Yi Jin Dan

In this chapter, we introduce breakthrough Divine treasures to serve humanity. These treasures are named Divine Shen Qi Jing He Yi Jin Dan. He means *join as*. Yi means *one*. Jin means *golden*. Dan means *light ball*. Divine Shen Qi Jing He Yi Jin Dan is a divine golden light ball of soul, heart, mind, energy, and matter joined as one. It can also be explained as a divine golden light ball of divine information, divine energy, and divine matter joined as one. The Divine is a creator. The Divine can create this golden light ball for any bodily system, organ, cell, cell unit, DNA, RNA, part of the body, and more. For example, the Divine can create and download a Divine Respiratory System Shen Qi Jing He Yi Jin Dan, a Divine Pancreas Shen Qi Jing He Yi Jin Dan, a Divine Breasts Shen Qi Jing He Yi Jin Dan, or a Divine Brain Cells Shen Qi Jing He Yi Jin Dan.

Now, the Divine will transmit as a gift to you, dear reader, one of these priceless, permanent Divine treasures, named:

Divine Heart Shen Qi Jing He Yi Jin Dan

We were given the authority and honor to offer a Divine Heart Shen Qi Jing He Yi Jin Dan to every reader. We connect with the Divine as we write this book. We asked the Divine, and the Divine agreed, to preprogram this transmission within these paragraphs. As long as you read this paragraph and are willing to receive this divine golden light ball, you will receive it shortly. You have free will. If you do not wish to receive this gift, simply tell the Divine that you are not ready. You will not receive this treasure. No one is forced to receive this treasure. We can assure you, though, that there are no negative effects from receiving this treasure. It is a karma-free treasure created in the moment in the heart of the Divine. It carries only positive shen qi jing.

If you are willing to receive, relax and sit up straight, with the greatest gratitude and honor in your heart. Silently speak as follows from your heart to the Divine:

> *Dear Divine,*
>
> *I am extremely honored to receive this gift of your permanent treasure, Divine Heart Shen Qi Jing He Yi Jin Dan, which is a divine golden light ball of divine heart information, energy, and matter joined as one. This means that you accumulate your heart information, energy, and matter to form a golden light ball to be transmitted to my heart. The divine heart's positive information, energy, and matter can transform the negative information, energy, and matter of my heart. If I practice with this treasure, I could receive healing for any heart issues. I could prevent sickness in my heart. I could even rejuvenate and prolong the life of my heart. I am extremely grateful.*

Prepare! Close your eyes and be silent for one minute.

Divine Heart Shen Qi Jing He Yi Jin Dan

Transmission!

Congratulations! You have received this permanent treasure, which will always remain with your heart and you. Use it well. The potential benefits you could receive are great.

Practice to Heal, Prevent Sickness, Rejuvenate, and Prolong the Life of Your Heart

How do you use your Divine Heart Shen Qi Jing He Yi Jin Dan?

Apply the Four Power Techniques:

Body Power. Put one palm on your lower abdomen, below your navel. Put your other palm over your heart.

Soul Power. Say *hello* to inner souls:

Dear shen qi jing of my heart and my heart cells,
I love you, honor you, and appreciate you.
You have the power to heal and rejuvenate yourselves.
Do a great job!
Thank you.

Dear my Divine Heart Shen Qi Jing He Yi Jin Dan,
I love you, honor you, and appreciate you.
You have the power to heal, prevent sickness, rejuvenate, and prolong the life of my heart.
Please bless my heart.
Thank you!

Mind Power. Visualize a divine golden light ball rotating, vibrating, and shining in your heart and all of your heart cells.

Sound Power. Chant repeatedly, silently or aloud:

> *Divine Heart Shen Qi Jing He Yi Jin Dan heal, prevent sickness, rejuvenate, and prolong the life of my heart.*
> *Divine Heart Shen Qi Jing He Yi Jin Dan heal, prevent sickness, rejuvenate, and prolong the life of my heart.*
> *Divine Heart Shen Qi Jing He Yi Jin Dan heal, prevent sickness, rejuvenate, and prolong life of my heart.*
> *Divine Heart Shen Qi Jing He Yi Jin Dan heal, prevent sickness, rejuvenate, and prolong the life of my heart. …*

Chant for at least ten minutes per time. You can adapt this practice to bring positive information to any aspect of your life. For chronic and life-threatening conditions, or for great challenges in relationships, finances, or any aspect of life, chant for two hours or more a day. Add all of your practice times together to reach two hours or more.

The longer you practice and chant, the more transformation you could achieve. The Divine Heart Shen Qi Jing He Yi Jin Dan carries divine positive information, energy, and matter, which is negative entropy. It takes time to transform the negative information, energy, and matter of serious challenges for health, relationship, finances, and other aspects in life. Negative information is negative karma. As with Walter and his liver cancer, Divine treasures can gradually clear negative karma, which is to transform entropy into negative entropy. With Divine and Tao treasures, we have created thousands of heart-touching and heart-moving results since 2003. Witness more than one thousand recorded stories on my YouTube channel.

Importance of Divine and Tao Treasures

I started to offer Divine treasures to humanity in 2003. In 2008, I was given the honor and authority from Tao to offer Tao frequency treasures to humanity. These treasures offer breakthrough healing, blessing, and transformation to humanity.

Everyone and everything is made of shen qi jing. Everyone's and everything's shen qi jing is in different layers of positivity and negative entropy. As human beings, we can hardly compare our shen qi jing with Divine and Tao shen qi jing.

As we have shared, the negative information, energy, and matter—the entropy—in our human shen qi jing causes disharmony and disconnection, internally and externally. This is why we get old, get sick, have physical, emotional, mental, and spiritual pain and imbalances, have troubled relationships, and lack abundance, flourishing, and true happiness and joy in our finances, intelligence, careers, wisdom, and every aspect of our lives.

Divine and Tao downloads and transmissions bring Divine and Tao shen qi jing to you. Divine and Tao are creators. They can create Divine and Tao shen qi jing for organs (as you have been offered above for your heart), bodily systems, tissues, and other parts of the body, including cells, cell units, DNA, RNA, tiny matter, and spaces. They can create Divine and Tao shen qi jing for relationships, finances, intelligence, and spiritual communication channels. They can create Divine and Tao shen qi jing for a pet, a house, and a business. They can create Divine and Tao shen qi jing for loved ones who have transitioned. They can create anything we can imagine. They can create anything beyond what we can imagine.

Divine and Tao shen qi jing downloads and transmissions can transform all kinds of negative information, energy, and matter to positive information, energy, and matter.

The Divine Heart Shen Qi Jing He Yi Jin Dan includes three light beings: Divine Heart Shen Jin Dan, Divine Heart Qi Jin Dan, and Divine Heart Jing Jin Dan. These three golden light balls join as one golden light ball when they are transmitted to the recipient.

Since July 2003, I have offered hundreds of thousands of Divine and Tao downloads and transmissions of systems, organs, parts of the body, and more. These Divine and Tao treasures have created hundreds of thousands of soul healing miracles.

Practicing with your Divine and Tao transmissions is vital to receive the greatest benefits. Do the practices in this chapter and later chapters as much as you can.

Practice, practice, practice.

To view thousands of soul healing videos, I welcome you to visit my YouTube channel, www.YouTube.com/zhigangsha. You could also receive great benefits from reading my book *Divine Soul Mind Body Healing and Transmission System: The Divine Way to Heal You, Humanity, Mother Earth, and All Universes.*[9] In this book, I have explained all kinds of shen qi jing blockages in detail. I have explained that soul blockages

---

[9] New York/Toronto: Atria Books/Heaven's Library Publication Corp., 2009.

are all kinds of negative karma. Mind blockages include negative mind-sets, negative beliefs, negative attitudes, ego, attachments, and more. Body blockages are energy (qi) blockages and matter (jing) blockages.

I have created almost one hundred fifty Divine and Tao servants. Appointed ones can offer Divine and/or Tao Shen Qi Jing He Yi treasures. Tao Shen Qi Jing He Yi treasures will continue to be given for transforming health, relationships, finances, business, intelligence, success, spiritual channels, occupations, enlightenment, prevention of sickness, rejuvenation, longevity, and more.

Thank you, Divine and Tao Source, for giving me the authority and privilege to train Divine and Tao servants to offer priceless Divine and Tao Source treasures to humanity, animals, and beyond. We are extremely blessed and grateful.

9

Tao Calligraphy

CALLIGRAPHY IS A REVERED ART in many countries and cultures. In China, it has been one of the most highly respected and revered artistic media for many centuries. Written Chinese is partly pictographic, partly ideographic, and more. It is not constructed with an alphabet. Chinese calligraphy brings grace, fluidity, energy, beauty, power, and more to the characters and phrases.

Yi Bi Zi

Tao Calligraphy is a special, unique form of Chinese calligraphy that I (Master Sha) created. The style of calligraphy underlying Tao Calligraphy is called Yi Bi Zi (一笔字). Yi means *one*. Bi means *stroke*. Zi means *word* or *character*. Yi Bi Zi is *one-stroke writing*.

English words are constructed from an alphabet of twenty-six letters. Chinese characters are traditionally constructed from sixteen types of individual strokes. These sixteen types of strokes include heng, a horizontal stroke; shu, a vertical stroke; pie, a left-falling stroke; na, a right-falling stroke; dian, a dot; gou, a horizontal hook; zhe, a right turn, and more. See figure 7 on the next page.

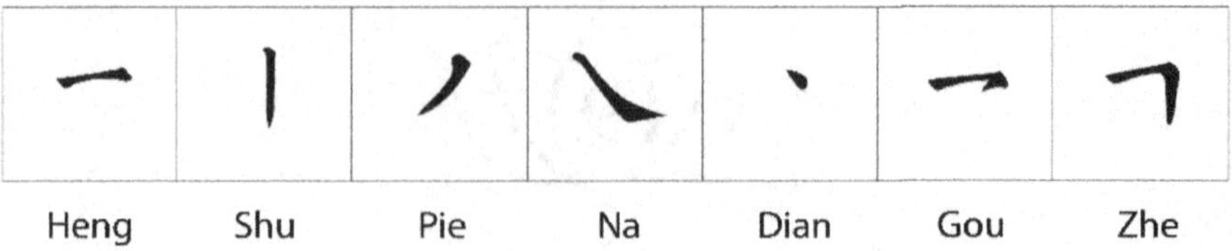

Figure 7. Examples of strokes in Chinese characters

A Chinese character consists of one or more such strokes. The only character written with a single stroke is "yi" (一, *one*). Some characters are written with more than thirty individual strokes. In Yi Bi Zi, every character is written with one continuous stroke, with the brush always in contact with the paper. Take, for example, the character 靈 ("ling," which means *soul*). As you can see, this character is rather complicated. To write it normally requires twenty-four individual strokes. Figure 8 shows "ling" written in Yi Bi Zi.

Figure 8. Ling (*soul*) in Yi Bi Zi

See and feel the free-flowing fluidity, the roundedness, the balance, the yin-yang alternation in the density, the one qi, and more of this special calligraphic style. Yi Bi Zi is Oneness writing, wherein every character, and sometimes entire phrases of multiple characters, is written in oneness.

What Is Tao Calligraphy?

Tao Calligraphy takes Yi Bi Zi to a special and unique level.

Recall Lao Zi's profound wisdom of Tao Normal Creation and the new Tao Science wisdom of Tao Reverse Creation. Tao Normal Creation is to go from Tao to wan wu (all things). Tao Source Oneness creates countless planets, stars, galaxies, universes, and human beings. Tao Reverse Creation is to return everyone and everything back to Tao Source Oneness.

In ancient wisdom, there is a profound sacred statement, Yi Zi Yi Tai Ji. Yi means *one.* Zi means *word.* Yi again means *one.* Tai Ji is a circle; this circle is also *one.* In one sentence:

Yi Ji Shi Yuan, Yuan Yang Zhen Qi

Yi is *one.* Ji Shi means *is.* Yuan means *circle.* Yang means *positive.* Zhen means *Tao,* the Source. Qi means *energy.* This means:

One is a circle, and the circle represents
Tao Source positive energy.

Two things make Tao Calligraphy unique and very special. Tao Calligraphy is the highest Oneness writing because:

1. It is Tao Yuan (Circle) Calligraphy. Tao Calligraphy is a special form of Yi Bi Zi that is more curvilinear and less angular than is typical. When we write Tao

Calligraphy, we literally make many turns—clockwise and counterclockwise circles—with the brush. The circle of Tao Normal Creation and Tao Reverse Creation tells us that Tao is the Source of everyone and everything, and that Tao holds everyone and everything to bring everyone and everything back to Tao. Therefore, the circle is One. The circle is Tao. This alone gives Tao Calligraphy the energy of Tao Oneness.

2. When writing a Tao Calligraphy, I literally transmit Tao Source shen qi jing (information, energy, matter) to the calligraphy. Simply to be in the presence of a Tao Calligraphy is to be in a Tao Source Oneness field (Tao Chang 道场—chang means *field*). When you connect with a Tao Calligraphy by doing simple specific practices that you will learn and experience in this chapter, the benefits you can receive from the positive shen qi jing of the calligraphy can be dramatic, phenomenal, and beyond imagination.

In Tao Science, everyone and everything is a vibrational field consisting of information, energy, and matter. All kinds of challenges, including sickness, relationship challenges, financial lack, lack of clarity of thinking and intelligence, and blockages in any aspect of life, are all due to a field that carries negative information, energy, and matter, which is entropy. Tao information, energy, and matter has the highest negative entropy, which is the most positive and purest information, energy, and matter. A Tao Calligraphy is a Tao Chang (*Tao field*) with Tao information, energy, and matter. Its power to transform negative information, energy, and matter to positive information, energy, and matter has no equal on Mother Earth. To have this Tao Chang available to us on Mother Earth is a blessing beyond words, comprehension, and imagination.

Receive Positive Information, Energy, and Matter from a Tao Calligraphy Tao Chang to Transform All Life

I have created a Tao Calligraphy especially for this book. See figure 9 following page 166. This Tao Calligraphy is one of the most significant and powerful messages of Tao Science:

Shen Qi Jing He Yi

Shen means *information*, which includes soul (content of information), heart (receiver of information), and mind (processor of information). Qi means *energy*. Jing means *matter*. He means *join as*. Yi means *one*.

Shen Qi Jing He Yi means:

Information, energy, and matter join as one.

You can use this Tao Calligraphy to transform all life. Here is how to do it.

Use the Five Power Techniques:

Body Power. Sit up straight with your feet flat on the floor and your back free and clear. Do not lean against the back of your chair. You may also stand with your feet shoulder-width apart.

Soul Power. Say *hello* to inner souls:

> *Dear all of my shen qi jing,*
> *I love you, honor you, and appreciate you.*
> *You have the power to heal, transform, and enlighten every aspect of my life.*
> *Do an amazing job!*
> *Thank you.*

Say *hello* to outer souls:

Dear Tao Calligraphy Shen Qi Jing He Yi,
I love you, honor you, and appreciate you.
You have the power to heal, transform, and enlighten every aspect of my life.
Please bless me. (Mention any blessing you wish for your health, relationships, finances, intelligence, success, enlightenment, and more.)
Thank you from my heart and soul.

Dear Tao Source,
Dear the Divine,
Dear countless planets, stars, galaxies, and universes,
I love you, honor you, and appreciate you.
You have the power to heal, transform, and enlighten every aspect of my life.
Please bless me. (Repeat your request for blessing.)
Thank you from my heart and soul.

Sound Power. Chant silently or aloud:

Shen Qi Jing He Yi (pronounced *shun chee jing huh yee*)
Shen Qi Jing He Yi
Shen Qi Jing He Yi
Shen Qi Jing He Yi

Information, energy, and matter join as one.
Information, energy, and matter join as one.
Information, energy, and matter join as one.
Information, energy, and matter join as one.

Shen Qi Jing He Yi
Shen Qi Jing He Yi
Shen Qi Jing He Yi
Shen Qi Jing He Yi

Information, energy, and matter join as one.
Information, energy, and matter join as one.
Information, energy, and matter join as one.
Information, energy, and matter join as one. ...

Chant for at least ten minutes per time. For serious, chronic, or even life-threatening health issues or challenges in any other aspect of life, including relationships and finances, chant for two hours or more each day. The more you chant, the more the positive information, energy, and matter of the Tao Calligraphy Tao Chang will transform your negative information, energy, and matter. The more you chant, the more your entropy will be transformed to negative entropy.

Mind Power. Visualize golden light shining in the area of your request. For example, for physical health, visualize the area or part of the body for which you requested the blessing. For a relationship, visualize yourself and the other person. For finances, visualize a golden money tree in your heart chakra. We emphasize this sacred ancient spiritual wisdom and power from Tao Source: *Golden light shines, all sickness and all challenges in life are transforming.*

Tracing Power. When you trace a Tao Calligraphy, you are connecting with its shen qi jing in a powerful way. We have already connected with the Tao Calligraphy *Shen Qi Jing He Yi* (figure 9) by "saying *hello*" as part of the Soul Power technique above. However, with Tracing Power you will receive much greater and faster nourishment and transformation from the positive information, energy, and matter within the calligraphy. Adding Tracing Power will multiply and accelerate the benefits by a factor of fifty or more! Pay great attention to Tracing Power.

Prepare. Put the fingers of one hand together. See figure 10. You will trace the calligraphy with your fingertips.

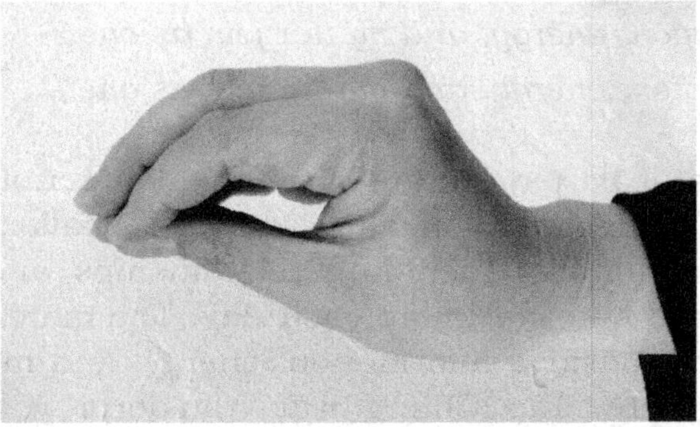

Figure 10. Five Fingers Tracing Power Hand Position

Trace the Tao Calligraphy *Shen Qi Jing He Yi* by following the path of the one-stroke writing. See figure 11 following page 166.

Trace and chant together for at least ten minutes. There is no time limit. The longer you practice and the more often you practice, the better the results you could receive.

The Five Power Techniques are very powerful. Body Power, Sound Power, and Mind Power have been applied since ancient times. To apply one technique is powerful. Soul Power was only known and used secretly by high-level spiritual masters and healers. Tao Calligraphy Tracing Power is possible only since I began to release Tao Calligraphy in my book *Soul Healing Miracles*[10] in 2013. To add Soul Power and Tao Calligraphy Tracing Power and apply the Five Power Techniques together is beyond extraordinarily powerful.

As of July 2017, I have trained hundreds of Tao Calligraphy Practitioners and twenty Tao Calligraphy Teachers worldwide to spread the positive information, energy, and matter of Tao Calligraphy.

Above all, I have created eight special Tao Calligraphy Tao Chang in my Tao Centers around the world. They are in Antwerp (Belgium), Toronto (Canada), Amersfoort (Netherlands), San Francisco (USA), Sydney (Australia), Vancouver (Canada), Honolulu (USA), and London (England). Each of these eight Tao Calligraphy Tao Chang holds forty to sixty-five special Tao Yuan (Circle) Tao Calligraphies to create the highest and most powerful Tao Source fields on Mother Earth. These Tao Calligraphy Tao Chang carry Tao infinite positive information, energy, and matter. They also carry the highest and purest negative entropy. These Tao Chang can transform all negative information, energy, and matter, as well as transform entropy into negative entropy. They offer breakthrough service to humanity through the principles of Tao Science.

We honor modern medicine. I myself am an M.D. from China. However, there are many sicknesses for which modern medicine and other healing modalities have no solution. The wisdom of Tao Science can explain this phenomenon. Simply, modern science and modern medicine have not realized enough the significance of soul issues, which are information

---

[10] *Soul Healing Miracles: Ancient and New Sacred Wisdom, Knowledge, and Practical Techniques for Healing the Spiritual, Mental, Emotional, and Physical Bodies.* Dallas/New York: BenBella Books/Heaven's Library Publication Corp., 2013.

issues. Information is divided into positive and negative. Negative information is the root cause of all sickness and all challenges—in relationships, business, success, intelligence, enlightenment, and every aspect of life.

Modern medicine does not believe in karma, which is a soul issue. Modern science studies information but it does not connect information with karma. In Tao Science, we explain scientifically that karma is information, and information is karma.

If we realize that negative information, energy, matter are the root cause of all sickness, of challenges in relationships and finances, and of all kinds of blockages to success in every aspect of life, we will focus on removing negative information, energy, and matter. In other words, to transform all life is to transform entropy into negative entropy.

We emphasize again: Tao Calligraphy is Tao Source Oneness writing. A Tao Calligraphy is a Tao field by itself. When you trace a Tao Calligraphy, the positive information, energy, and matter (or, in other words, the negative entropy) within the calligraphy will come to your body and to your life to transform your negative information, energy, matter, and entropy.

You can also learn to write Tao Calligraphy. My Tao Calligraphy Teachers offer courses worldwide. We welcome you to learn this Tao Oneness writing to benefit every aspect of your life.

In a few short years, Tao Calligraphy has created thousands of amazing results for healing and life transformation. It will serve much more. It brings Tao Source positive information, energy, matter, and negative entropy to every aspect of life. Realize in your heart and soul this profound wisdom and practice.

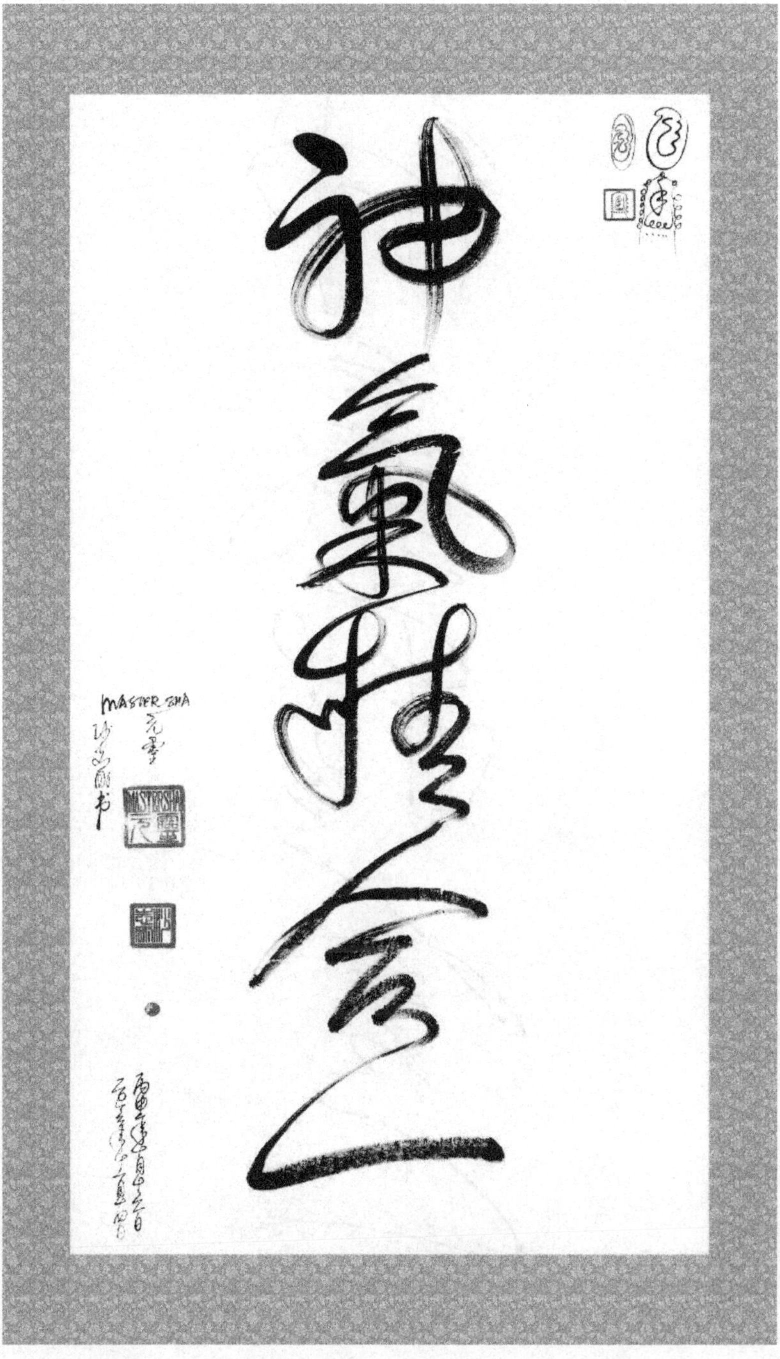

Figure 9. Tao Calligraphy *Shen Qi Jing He Yi*

Figure 11. Tracing pathway of *Shen Qi Jing He Yi*

10

Law of Karma

KARMA IS A FUNDAMENTAL TEACHING in many spiritual belief systems. Millions of people believe in karma. Millions of people do not believe in karma. We feel this is the right time to explain karma in a scientific way. We are delighted that Tao Science can explain karma in a scientific way. The Law of Karma is a natural law.

Have you ever wondered why your life goes smoothly in some ways but has challenges in other aspects? Have you wondered why you repeat certain patterns that seem to follow you in your life, wherever you go and whatever you do?

Do you know why some people keep making the same mistakes over and over again, no matter how much advice they receive to change these behaviors, and even when they know they should change?

Do you know why you like some people you meet instantly, while others irritate you when you meet them, just by their appearance or the sound of their voice?

All of these actions and responses are due to karma. What is karma? Karma is the record of services we and our ancestors have offered in this lifetime and all past lifetimes. Our karma has great power over us. Our karma can make us think, feel, hear, speak, write, smell, taste, eat, react, behave, make decisions, and experience life in certain ways. Our karma can literally control us, blind us, deafen us, bind us, imprison us, and enslave us to certain negative karmic patterns.

What is the root cause of our health and relationship issues, emotional and mental problems, financial challenges, and many other difficulties in our lives? What is the root cause of natural disasters, human catastrophes, violence, tragedies, energy and financial crises, pollution of earth, air, water, and food, global warming, and many other challenges in our societies and world? The root cause is negative karma. Negative karma is the hurtful, harmful, selfish actions, behaviors, words, and even thoughts that we and our ancestors have created in all of our lifetimes. The complete record of our past actions, behaviors, words, and thoughts is stored in our vibrational field. This Akashic record—our karma—determines every major aspect of our lives. To heal ourselves, our societies, and the world, we need to clear our negative karma.

The Law of Karma, which we present in this chapter, has been known for thousands of years to many spiritual traditions, cultures, and disciplines. It is the foundation of many spiritual teachings. The wisdom of the Law of Karma has helped relieve much suffering for humanity. At this critical historic moment, it is crucial for us to understand the Law of Karma in both scientific and spiritual ways. It is urgent for each one of us and all humankind to clear our negative karma.

Now, a scientific understanding of the Law of Karma is revealed to humanity. For the first time ever, the sacred wisdom, knowledge, techniques, and power about how to clear negative karma are brought to humanity. This wisdom, knowledge, techniques, and power are given to us now because they are urgently needed to help humanity and

Mother Earth pass through this difficult transition period. They are needed to literally save humanity and Mother Earth. Moreover, this wisdom empowers each of us with higher abilities to create and manifest. With this wisdom, each one of us and all humanity can be uplifted to a higher level of existence with deeper meaning to our lives, with more love, joy, abundance, beauty, health, and wisdom. We can create a world of love, peace, and harmony.

What Is Karma?

Karma is the record of our and our ancestors' services in all lifetimes, present and past. Karma is also called virtue or deed. Karma can be divided between good (positive) karma and bad (negative) karma. Good or positive karma is the record of good services, including love, forgiveness, compassion, light, generosity, kindness, purity, integrity, and more. Bad or negative karma is the record of unpleasant services, including killing, harming, taking advantage of others, stealing, and more. Negative karma is a spiritual debt. When you hurt or harm another being or entity, you are in debt to that being or entity spiritually. You owe that being or entity, and will have to repay that debt or somehow have that debt erased in another way.

Good karma is measured by virtue. Virtue is spiritual money. Virtue can be measured by a high-level spiritual being. Advanced spiritual beings with powerful Third Eyes can see that good virtue is given as beautiful dots and flowers of different sizes from Heaven's virtue bank. These dots and flowers can be red, golden, rainbow-colored, purple, crystalline, and more. Ten small dots form a large dot. Ten large dots form a small flower. Ten small flowers form a large flower. Therefore, a flower represents more virtue than a dot. A big flower represents more virtue than a small flower. It is like various denominations of physical money on Mother

Earth, for example, ten- and twenty-five-cent coins and paper currency of one, five, ten, and twenty dollars. When you offer good service to humanity and others, you are given dots of virtue. Groups of dots will form flowers. If you offer great good service, you can be given different sizes of flowers directly. This virtue, expressed in dots and flowers, will come to your book in the Akashic Records and to your soul at the same time. When this virtue comes, the spiritual debt in your Akashic book is being paid down, little by little.

On Mother Earth, you may borrow money from a bank to purchase a house. You then have a physical debt to the bank. You must pay that debt little by little, month by month, according to the contract you have with the bank. It is common for the borrower to be obligated to pay this debt gradually for twenty, thirty, or forty years.

Bad karma is your spiritual debt. This spiritual debt is recorded in your book in the Akashic Records and on your soul. You may have made mistakes in your past lives and in this life. When you kill, steal, cheat, take advantage of, hurt, or harm others in any way, you create a spiritual debt. Spiritually, you owe the souls you have harmed. Just as you must pay your mortgage to the bank, you must pay your spiritual debt.

How do you pay this spiritual debt? When you have bad karma, you are given lessons to learn. These lessons could include sickness, accidents, broken relationships, financial challenges, mental disorders, emotional imbalances, and all kinds of blockages in life. When you are faced with a challenge in your life, when your family has a challenge, when you are seriously ill, or when you feel unhappy, stressed, angry, fearful, depressed, or anxious, you may not think about karma. We would like to share our insight that significant challenges in your physical body, emotional body, mental body, and spiritual body are almost always due to karmic issues. To pay your spiritual debt is to have lessons to learn.

The one-sentence secret about karma is:

Karma is the root cause of success and failure in every aspect of life.

Because bad karma is the root cause of blockages and challenges in life, we must clear (or cleanse, which is a different word for the same thing) our bad karma, which is our spiritual debt from all our lifetimes, in order to transform our lives.

People on Mother Earth suffer so much. For the physical body alone, there are thousands of sicknesses. Most people on Mother Earth have not realized that serious, chronic, and life-threatening health conditions are due to bad karma. Even less do people understand that all of the major blockages in their lives are due to bad karma.

To heal our physical, emotional, mental, and spiritual bodies, we have to learn how to self-clear our bad karma. To self-clear karma, we should first understand what kinds of bad karma human beings have.

Personal Karma

You could have had hundreds or even thousands of lifetimes as a human being. Everyone makes mistakes. In some lifetimes, you could have made huge mistakes. For example, in one lifetime you may have been the leader of a country or an important general. You could have forgotten to offer love, care, and compassion. You may have harmed others, caused a war, or even been directly involved in killing others. You could have been a very wealthy and influential person who took advantage of others.

Heaven records these actions and behaviors. The harm that you caused others creates your spiritual debt. This harm is arranged to return to you in some form in your current life and your future lives. These are the lessons you need to learn.

Let us give you an example. In Japan, I (Master Sha) had a consultation with a woman who told me she was very upset because her husband had six girlfriends. I asked her, “Do you believe in karma?” She replied, “Yes. I believe in karma.” I then asked, “Do you want to know your karmic issues related to this situation with your husband?” She replied, “Yes, please.”

I asked the woman to close her eyes. I connected with the Akashic Records and asked Heaven to show me the past-life relationships between her husband and her. In about twenty seconds, I got an answer. I told her to open her eyes and said, “Your problem with your husband is due to you.” She was surprised and said, “Really? It is because of me?” I explained, “Yes. I did a spiritual reading with the Akashic Records. Heaven showed me that in a past life, you were also married to your husband. You were the husband and your current husband was your wife. In that lifetime, you had more than twelve girlfriends. Heaven showed them to me one by one in my Third Eye.”

My client was stunned, but she understood. She asked me what she should do.

I told her, “Forgive your husband. Give him love. Love melts all blockages and transforms all life.”

This example is to tell you to remember that if you have significant relationship challenges, there is almost certainly a spiritual reason behind them. If someone is hurting you, you could have created negative relationship karma in your previous lifetimes together. You may have hurt the other person. The hurt you are receiving now is your spiritual lesson to pay your spiritual debt.

We often focus on health when we speak or think about healing. In fact, healing is needed in every aspect of our lives. You may have relationship challenges, emotional challenges, work or school challenges, financial challenges, and more. All of this suffering needs healing. To heal all kinds of challenges effectively, it is vital to understand deeply that any and all major challenges in your life could be due to negative karma. Karma is the root cause of success and failure in every aspect of life.

Ancestral Karma

Everyone has two parents. Everyone has four grandparents. Everyone has eight great-grandparents. Our ancestral tree extends back in time for hundreds and thousands of generations. We have a huge number of ancestors in this lifetime. In addition, you could have had hundreds or thousands of previous lifetimes as a human being. Over all of these lifetimes, you probably had many different fathers and many different mothers. Your combined ancestral tree from all your lifetimes could include millions and millions of ancestors. There is a renowned ancient saying:

Qian Ren Zai Shu, Hou Ren Cheng Liang

This can be translated as: *Ancestors plant the tree, descendants enjoy the shade.*

This saying is part of the wisdom of the Law of Karma. It says that ancestral karma exists, and that we all carry it. Karma, positive and negative, is passed from one generation to the next. In this way, you receive some good ancestral karma, and the blessings that come with it, from the good services your ancestors offered. On the other hand, you receive some bad ancestral karma, and the lessons it brings, from the hurtful and harmful services your ancestors offered. It is like inheriting physical characteristics from your parents and grandparents. The Law of Karma says that we also inherit spiritual characteristics from our ancestors.

Because everyone and everything are made of shen qi jing, there is a multitude of kinds of karma, including for example relationship karma, financial karma, emotional karma, mental karma, speaking karma, spiritual communication karma, spiritual abilities karma, writing karma, science and research karma, and much more. In one sentence, every aspect of life is related with karma. Karma wisdom is vital to understand and transform life.

Dr. Rulin Xiu's Journey to Understand Karma Scientifically

"Do you believe in karma? Raise your hand if you believe in karma."

On the fateful day, September 9, 2009, I (Rulin Xiu) met Dr. and Master Sha for the first time, attending his one-day workshop in my neighborhood in Hawaii. He put this significant question to all one hundred of us in the workshop.

I raised my hand in response. At the same time, as a Berkeley-trained theoretical physicist, I started to think deeply in my heart and mind about how to understand karma from the perspective of fundamental physics.

Can we derive the Law of Karma, this important spiritual law, from theories of physics? Can we express the Law of Karma mathematically? In other words, is the Law of Karma also a physical law?

Master Sha is very passionate to teach humanity about karma. As we were writing this book, a message came to me: Master Sha has come to Mother Earth at this critical time to teach humanity about karma. Much more, he is here to help humanity remove negative karma. I was shown a spiritual image of Noah on his ark with most of humanity having been wiped out from great floods and other disasters. I had a sudden realization that it is critical for each of us and humanity to know about karma and to clear our negative karma at this historic moment. Otherwise, our negative karma could cause huge damage for humanity and Mother Earth—and ourselves.

When I mention karma to people, most people have a negative feeling because they may think about the consequences and lessons from past mistakes as punishment they must endure. However, as I have come to understand the Law of Karma scientifically, I have realized that the Law of Karma is the most empowering physical and spiritual law. It tells us that we are the creator of our own reality. It shows us how we can manifest anything and everything we could ever want. It reveals to us that the unlimited freedom and miraculous abilities we can achieve are beyond imagination. The Law of Karma holds the key for liberating humanity from suffering. It is the gateway to uplift everyone and all humanity to a higher level of existence.

There is never anyone or anything to blame. There is no need to be anxious or stressed. All is in our own hands right now, right here. We are in control of everything in our own reality. We can manifest anything we would ever want.

To become a powerful creator and to be in ultimate control of your reality, you need to learn the Law of Karma.

Law of Karma Expressed Mathematically

Many people think of Newton's Third Law of Motion as the scientific expression of the Law of Karma. Sir Isaac Newton (1642–1726) was the founder of classical physics. Everyone has heard the story about his discovery of gravity when an apple fell from the tree above him onto his head. Newton formulated the three fundamental laws of motion, establishing the foundation for classical mechanics. Newton's Third Law of Motion states:

For every action, there is an equal and opposite reaction.

Newton's Third Law tells us that what we give to others is what we receive in return—in exactly the same way and the same degree. It aligns with the Golden Rule:

Do unto others as you would have others do unto you.

Newton's Third Law of Motion conveys the main idea of the Law of Karma: what we give to others is what we receive from others. It teaches us that if we want to receive something, we need to start with giving the same thing to others.

Newton's Third Law does not tell us how our actions and the actions of our ancestors affect our thoughts, feelings, relationships, finances, and every aspect of our lives. For example, suppose we try to interpret Newton's Third Law of Motion literally. If we give someone an apple, we should get another apple back at the same time. If we shoot a bullet at someone, there should be another bullet coming at us instantly. Obviously, this is not what happens in real life. Real life is much more complicated than this. Newton's Third Law of Motion, as well as classical physics, is too simplistic to describe the Law of Karma in its full subtlety.

To understand the Law of Karma scientifically, we need quantum physics and Tao Science.

Karma and Vibrational Field

Quantum physics reveals to us that everyone and everything is made of a vibrational field. Our vibrational field consists of various vibrations with polychromatic frequencies, wavelengths, and other properties. Our vibrational field carries all the information, energy, and matter about us. This vibrational field is the record of our actions, as well as the impact on us from the actions of our ancestors, humanity, Mother Earth, the solar system, countless galaxies, and countless universes.

Our vibrational field expands over all space and time. It is not limited in any way. It is part of a universal vibrational field. The universal vibrational field contains everyone and everything. Each of our actions affects the entire universal vibrational field. It affects everyone and everything. It is a service to everyone and everything.

Karma is the record of services.

Karma is the information, energy, and matter carried by our vibrational field about the past services offered by our ancestors and ourselves.

Karma is recorded within the vibrational field. Our vibrational field records all of our actions, behaviors, and thoughts from the beginning of our existence. Our vibrational field is our Akashic record, from which we can read our karma.

Karma can be divided into good karma and bad karma. We have given the spiritual definition for good karma and bad karma. Now let's give them a scientific definition.

Scientific Definition of Karma

We can define karma scientifically, in terms of information and entropy. Let's start with good karma.

Good karma is action that increases positive information. Good karma can be measured by the increase in negative entropy.

Positive information is the measure of connection with all beings. Action that enhances our quantum entanglement and other connections with all beings increases positive information. The more connection or positive information we have, the more power, wisdom, and influence we have.

Good karma builds connection. Good karma enhances our soul power, heart power, and mind power. It leads to love, joy, abundance, longevity, wisdom, and peace. Love, forgiveness, compassion, light, humility, harmony, prosperity, gratitude, service, and enlightenment can increase our good karma.

The mathematical measure of good karma is negative entropy. Negative entropy is the mathematical measure of virtue. The mathematical measure of negative entropy is similar to the spiritual measure of virtue.

Now, the scientific definition of negative karma should be no surprise.

Negative karma is action that increases negative information. Negative karma can be measured by the increase in entropy.

Negative karma increases our negative information. Negative information is the disorder and uncertainty within ourselves and our disconnection from others. The less connection we have with others, the less power, wisdom, and influence we have. Lack of connection is expressed by the negative information within us. The negative information is measured by entropy. Entropy expresses how much disorder, disconnection, and uncertainty we have. Entropy is the mathematical measure of karmic debts.

Negative karma disconnects us from all beings. It reduces the power of our soul. It closes our heart, limiting its ability to receive. It limits our mind.

Killing, stealing, cheating, taking advantage of others, judging, discriminating, complaining, and more are unpleasant services that decrease our connection with others. They generate negative information. Negative information or entropy increases the disorder within us. It leads to difficulties, challenges, sickness, disasters, decay, and death.

To reduce negative karma is to conduct action that restores and strengthens connection with others. It is to increase our positive information, which is virtue. Our positive information, virtue, will offset and reduce our spiritual debts, which are negative karma.

What causes negative karma? Buddha teaches us that greed, anger, and ignorance can lead to negative karma. Greed, anger, and ignorance lead to separation and disconnection from others. All kinds of judgment, discrimination, and attachment can also create negative karma. To avoid creating negative karma, we need to remove greed, anger, ignorance, judgment, discrimination, attachment, and more.

Dear reader, at this moment, we ask you to pause and take some time to examine your thoughts, your speech, your feelings, your emotions, your intentions, your hearing, your seeing, your writing, your reading, and each of your actions.

Are you creating positive karma that will make you happier, healthier, and more successful? Or, are you creating negative karma that will bring you unhappiness, sickness, difficulties, disasters, challenges, and other negative experiences in your health, relationships, finances, careers, and more?

Law of Karma

The Law of Karma expresses how our actions manifest our reality. The Law of Karma has two parts. The first part is as follows:

Law of Karma, Part 1

Part 1 of the Law of Karma explains what determines our experiences in every aspect of our lives. This is the core of the Law of Karma.

What we are experiencing right now in our spiritual, mental, emotional, and physical bodies, as well as in our relationships, finances, and every aspect of our lives, is what our ancestors and we have given to others in the past.

Scientific Derivation and Explanation of the Law of Karma, Part 1

The first part of the Law of Karma tells us how our past actions affect our current reality. Each of our actions, such as thinking, speaking, feeling, hearing, smelling, tasting, emoting, moving, and more, manifests a set of vibrations. Let's suppose you have acted in some way. Let's call that action "A." Action A manifests a set of vibrations, called "B." Because B comes from you, B is quantum entangled with your vibrational field.

Now let's suppose a person named John receives B. John experiences happiness when he feels B. The action of John experiencing B as happiness is the manifestation of B as a state of happiness.

Since B is quantum entangled with part of your vibrational field, part of your vibrational field will now move into a state of happiness. If your heart receives these "happy state" vibrations, you will feel happiness. Therefore, because your action made John feel happy, you also experience happiness.

Suppose on the other hand that John does not immediately receive B, but he experiences it at a later time. In this case, B will manifest as a state of happiness at a later time. This happy state will then come into your vibrational field at a later time as well. In other words, you will receive the karmic consequence of action A at a later time. This shows us that karmic consequences can be delayed.

In either situation, the vibrations in the happy state caused by A exist. John may receive these vibrations immediately or at a later time.

Now suppose your heart does not receive the vibration in the happy state. If this is the case, even if it exists in your vibrational field, you still won't feel the happiness caused by your action A. When the time is right for you to receive this happy vibrational field, you will experience the happiness caused by your own action A. This is another way that karmic consequences can be deferred.

If many people receive B and experience happiness, you will experience more happiness. The more people experience B as happiness, the more happiness you will experience. The longer and more often others feel happiness from B, the longer and more often you may feel happiness from your action A.

If, for whatever reason, some people experience sadness when they feel B, you may also feel sadness. You may experience both happiness and sadness from your action A.

From these examples, we can see that our karma can affect us for a long time and in many ways.

Our karma can affect our children. This is because our children's vibrational field is partially quantum entangled with ours. Therefore, what we make others experience can also affect our children. It may affect them for a long time. In the same way, our ancestors' karma can affect us for a long time.

From the above examples, we can see that our karma can affect us for a long time and in many ways. Karma has four important characteristics:

1. The effects of karma can be delayed.
2. Karma has many layers.
3. Karma has a cumulative effect.
4. Karma is inherited.

If our action increases overall connection and order, it enhances positive information. We create positive karma. Positive karma is measured by negative entropy. With the creation of positive karma, our negative entropy increases. As we have mentioned before, positive karma, positive information, more negative entropy will make us healthier, younger, wiser, more powerful, and more successful.

If our action brings overall disconnection and disorder, it will enhance our negative information. We produce negative karma. Negative karma is measured by entropy. Negative karma, which is negative information and more entropy, will make us sick and older, and bring suffering, difficulties, challenges, failures, disasters, and death.

Each of us has many kinds of karma stored in our vibrational field. We will experience our karma when the time is ready. When is it time for us to experience our karma?

Karmic consequences happen in two ways: a positive way and a negative way. The positive way is due to positive karma, which is positive information measured by negative entropy. Positive karma will allow us to have more love, and be healthier, younger, wiser, more powerful, and more successful. When negative entropy accumulates and reaches a certain point, positive karmic consequences will occur. For instance, a person may be very poor. If the person serves diligently to help many others have a better life, his good deeds accumulate. At a certain point, the person will start to enjoy a better life.

The negative way is due to negative karma, which is negative information measured by entropy. Negative karma will bring sickness, aging, suffering, difficulties, challenges, failures, disasters, and death. When entropy accumulates and reaches a certain point, negative karmic consequences will happen. For example, a certain part of one's body has some inflammation. The inflammation worsens and spreads. The inflammation brings negative information, disorder, and disconnection to the body. When the inflammation grows to a certain point, serious sickness can occur.

We all want to have better health, youthfulness, success, wisdom, power, and prosperity in our lives. We all hope to reduce and eliminate sickness, aging, suffering, difficulties, challenges, failures, disasters, and death. To achieve this, we need to conduct good deeds to create more positive karma and increase our negative entropy. This negative entropy can reduce our entropy. In this way, our positive karma can compensate for our negative karma. Through this process, sickness, aging, suffering, difficulties, challenges, failures, disasters, and death can be prevented and eliminated.

Inferences from Part 1 of the Law of Karma

Inference 1

What we are receiving now is what we have given to others before.

Karma is the root cause of success and failure in every aspect of our lives.

Karma is cause and effect. When we give positive influences to others, we receive blessings. When we give negative influences to others, we experience challenges and learn lessons.

According to the Law of Karma, our current feeling, hearing, seeing, smelling, tasting, thinking, emoting, intelligence, finances, career, relationships, and every aspect of our lives are effects. The causes are our and our ancestors' past actions that have made others feel, hear, see, smell, taste, think, emote, and have certain intelligence, finances, careers, relationships, and more.

Inference 2

Karmic consequences appear accordingly.

All of our karma, positive and negative, is recorded in our vibrational field. Nothing is or will be missing. Some people may think they can hide what they or their ancestors did in the past. The truth is nothing can be hidden. There is an ancient statement: *if you don't want people to know, do not do it.* If you do it, you know. Heaven and Earth know also.

Whether or not we are aware of karmic consequences, we cannot avoid them.

There is a Chinese saying: Shan you shan bao, e you e bao. Bu shi bu bao, shi hou bu dao.

Shan means *kind.* You means *has.* Bao means *response* or *return.* E means *evil.* Bu means *not.* Shi means *is.* Shi hou means *time.* Dao means *arrive.*

This Chinese saying means: *Kind action has good returns. Harmful action incurs harmful returns. If you have not received the effect of your action, it is not because you will not be affected. It is because the time has not come.*

Inference 3

Karma may blind and trap us.

Our karma can blind us and control us. It can make us think, feel, see, hear, taste, smell, speak, and act in certain ways. It can make us believe that what we think and feel is the only truth. Many people are stuck in their particular point of view about life because of their karma. They are not open to higher wisdom and knowledge. This is one way karma traps us in life.

To live a life with more freedom, better health, and greater success in every aspect, it is crucial to know that our negative karma affects our feelings, ideas, thoughts, speaking, hearing, seeing, smelling, tasting, and more. To transform our lives, it is critical that we learn and follow higher wisdom.

Law of Karma, Part 2

The second part of the Law of Karma expresses how our reality is manifested. It tells us that our souls, hearts, and minds manifest our physical realities. The soul gives information to the heart. The heart receives the information and activates the mind. The mind processes the information and directs energy. Finally, the energy moves the matter. Matter is the physical reality we observe. The information your soul delivers, the ability of your heart to receive this information, and the way your mind processes the information determine what happens in your life.

What we have in our souls, hearts, minds, and bodies is what we will manifest in our physical realities.

In quantum physics, detectors are used to receive and process vibrations and exhibit quantum phenomena visibly. The kinds of vibrations that are present, the types of detectors we use, and where and how we place the detectors determine the phenomena we observe. Our souls give information. Our hearts and minds correspond to detectors in quantum physics. The kinds of information our souls give to our hearts, the information our hearts are able to receive, and the way our minds process the information determine every aspect of our lives.

The Law of Karma tells us not only how our past service determines our current reality, but also how our current actions shape our future. What is happening in our souls, hearts, and minds right now produces what will happen in our lives in the future. The choices we are making at this moment in our thoughts, feelings, emotions, attention, speech, writing, smelling, tasting, hearing, and more decide our future. Our choices profoundly affect our health, relationships, finances, and every aspect of our lives going forward.

Inferences from Part 2 of the Law of Karma

Inference 4

We are the creator of our reality.

Our fate is in our own hands. More strictly speaking, our realities are within our souls, hearts, minds, and bodies. We are the ones who are in control of our own realities and destinies. By changing and transforming our souls, hearts, minds, and bodies, we can transform our health, relationships, finances, intelligence, appearance, longevity, and every aspect of our lives.

Inference 5

At the present moment, and in every moment, we have the choice to create the life we want.

The Law of Karma is for us to realize that at the present moment we have choice. The choices we make at this moment decide what will happen in our lives in the next moment and the future.

At this moment, are we choosing anger or forgiveness and compassion? Are we in joy or worry? Are we loving or hating? Are we peaceful or fearful? Are we happy or sad? The choices we make in each moment determine our health, relationships, finances, and every aspect of our lives. If we keep worrying, we will create a life full of worry. If we focus on being sad, we will experience a sad life. If we are fearful, many things will come to our life to scare us. If we take advantage of others, things will be taken away from us. When we serve others, more will be given to us.

The Law of Karma tells us the power of our current actions. Although our current experiences are determined by what we have done in the past, at this moment, we also can choose a different way to think, feel, see, hear, taste, smell, speak, and act. By choosing a different way, we can and will change our futures. We can alter our destinies. We have the power to heal, transform, and uplift every aspect of our lives. To realize this is one of the highest liberations and empowerments.

How to Clear Negative Karma

Negative karma is our disconnection from others, as well as the disorder and uncertainty within us. It is the cause of sickness, aging, death, suffering, difficulties, and challenges in our health, relationships, finances, and every aspect of our lives. To remove negative karma is to remove all of this negative information. It is to gain more love, peace, power, wisdom, freedom, and abundance.

From our scientific understanding about karma, we find there are three ways to clear negative karma:

1. Transform negative information to positive information.
2. Apply forgiveness and offer service.
3. Remove darkness in our vibrational field.

Now let us see how we can remove our negative karma with these methods.

Transform negative information to positive information

Negative karma is the negative information within our souls, hearts, minds, and bodies. To transform the negative information to positive information is to clear our negative karma and create positive karma.

The Soul Light Era began on August 8, 2003. It will last at least fifteen thousand years. At the beginning of this era of "soul over matter," the Divine gave us a profound and powerful treasure to help us clear our negative karma. This sacred treasure is the Divine Soul Song, *Love, Peace and Harmony.*

This Divine Soul Song carries divine information and vibration. To sing *Love, Peace and Harmony* is one of the simplest and most powerful divine ways to self-clear your negative karma.

I (Master Sha) received *Love, Peace and Harmony* from the Divine on September 10, 2005. Its lyrics are very simple:

I love my heart and soul
I love all humanity
Join hearts and souls together
Love, peace and harmony
Love, peace and harmony

The first line, *I love my heart and soul,* is to purify our souls, hearts, minds, and bodies. Love melts all blockages and transforms all life. To love the heart and soul is to heal the heart and soul. Heal the heart and soul first, and then healing of the mind and body will follow. In Tao Science, love is a quantum field carrying positive information, energy, and matter that can transform our negative information, energy, and matter.

I love my heart and soul is a soul mantra to heal all of our sicknesses. It does take time to restore health from chronic and life-threatening conditions, but it definitely works. This soul mantra is powerful beyond comprehension.

The second line, *I love all humanity*, is to serve all humanity. In ancient spiritual wisdom, to transform negative karma, which is negative information, one has to serve. To serve is to make others happier and healthier. To give love is to serve. *I love all humanity* is to serve humanity. The more one serves, the more one increases positive information, and the more the negative information in our Akashic record will be released.

The third line, *Join hearts and souls together*, is a calling for humanity and all souls to align as one. This calling is the greatest service. Only through service can one's negative karma be forgiven.

The fourth line, *Love, peace and harmony*, is the ultimate goal we intend to achieve. A person needs love, peace, and harmony. A family needs love, peace, and harmony. An organization needs love, peace, and harmony. A city needs love, peace, and harmony. A country needs love, peace, and harmony. Mother Earth needs love, peace, and harmony. Countless planets, stars, galaxies, and universes need love, peace, and harmony.

The fourth line *Love, peace and harmony* is emphasized and repeated as the fifth and final line. It is a Divine Order to wan ling (*all souls*).

This Divine Soul Song, *Love, Peace and Harmony* is positive information, energy, and matter. If a fish lives in polluted water, the fish either gets sick or dies. To save the fish, purify the water. Humanity lives in a polluted environment on Mother Earth, including humanity's pollution of the soul, heart, and mind, environmental pollution of the air, water, land, and food, internal pollution in our own jing qi shen, and much more. To save humanity and Mother Earth, purify humanity's jing qi shen and Mother Earth's jing qi shen.

We emphasize that the blockages of jing qi shen we carry *are* pollution. Jing blockages are inside the cells. Qi blockages are in the space between cells. Shen blockages include blockages in soul, heart, and mind. Soul blockages include all kinds of negative karma, which is negative information in Tao Science, such as negative personal karma, negative ancestral karma, negative relationship karma, negative financial karma, and much more. Negative karma is negative information, which could exist in every aspect of life.

Why is the Divine Soul Song *Love, Peace and Harmony* extremely powerful? Love carries the most powerful positive information, energy, and matter, which can purify our negative information, energy, and matter in every aspect of life. Therefore, the Divine Soul Song *Love, Peace and Harmony* is the Divine's gift to empower us to clear our negative karma in every aspect of life. What we need to do is sing *Love, Peace and Harmony*. More than one million people have already received positive results by singing *Love, Peace and Harmony.*

In Tao Science, *Love, Peace and Harmony* creates positive quantum entanglement. The field of this Divine Soul Song can bring positive quantum entanglement with everyone and everything, everywhere. Therefore, the Divine Soul Song *Love, Peace and Harmony* is the Divine's treasure to clear our negative karma in order to transform our health, relationships, finances, and every aspect of life.

Sing the Divine Soul Song *Love, Peace and Harmony* to serve humanity and all souls. When millions of people chant this Divine Soul Song, great positive information will come to humanity and Mother Earth. This is very much needed because humanity is currently facing huge challenges, including environmental pollution, financial crises, wars and other conflicts, as well as other challenges in health, relationships, and more. These difficulties and issues are due to humanity's karma. If millions and billions of people were to chant this Divine Soul Song, humanity's karma would be reduced. This can help reduce air, water, and other pollutions. It can help end wars, famine, and other challenges.

I have founded the Love Peace Harmony Foundation, which is supported by the Love Peace Harmony Movement. The most important service and practice of the Love Peace Harmony Movement is to create a Love Peace Harmony World Family and a Love Peace Harmony Universal Family by singing *Love, Peace and Harmony.* I welcome you to get a free mp3 download of *Love, Peace and Harmony* at www.lovepeaceharmony.org. Singing the Divine Soul Song *Love, Peace and Harmony* for fifteen minutes a day could help you beyond comprehension. Join the Love Peace Harmony Movement to spread love, peace, and harmony to humanity and all souls.

Apply forgiveness and offer service

We teach Ten Da. Da means *greatest.* The Ten Da are ten greatest qualities of the Divine and Tao Source. The Ten Da are:

- Da Ai (*greatest love*)
- Da Kuan Shu (*greatest forgiveness*)
- Da Ci Bei (*greatest compassion*)
- Da Guang Ming (*greatest light*)
- Da Qian Bei (*greatest humility*)
- Da He Xie (*greatest harmony*)
- Da Chang Sheng (*greatest flourishing*)
- Da Gan En (*greatest gratitude*)
- Da Fu Wu (*greatest service*)
- Da Yuan Man (*greatest enlightenment*)

Ten Da is the nature of Tao Source, the ultimate Creator. Tao Source carries unlimited positive information, energy, and matter. As we will share in chapter eleven, Tao Source creates Heaven and Earth, which are yang and yin. Yin-yang interaction creates everything. We must realize deeply in our hearts and souls that Tao is the ultimate Creator.

Because Ten Da carries the highest positive information, energy, and matter, Ten Da can purify and transform every aspect of life. Let us practice now with two of the Ten Da to self-clear negative karma.

Da Kuan Shu (*Greatest Forgiveness*)

Da means *greatest.* Kuan Shu means *forgiveness.* Da Kuan Shu means *greatest forgiveness.* Forgiveness practice is the golden key to clear negative karma. Tao Source has given me four sacred phrases for each of the Ten Da. For Da Kuan Shu, the four phrases are:

Er Da Kuan Shu

Er means *two* or *second.* Da means *greatest.* Kuan shu means *forgiveness.* Er Da Kuan Shu means *the second of the Ten Da is greatest forgiveness.*

Wo Yuan Liang Ni

Wo means *I.* Yuan liang means *forgive.* Ni means *you.* Wo Yuan Liang Ni means *I forgive you.*

Ni Yuan Liang Wo

Ni means *you.* Yuan liang means *forgive.* Wo means *I.* Ni Yuan Liang Wo means *you forgive me.*

Xiang Ai Ping An He Xie

Xiang ai means *love.* Ping an means *peace.* He xie means *harmony.* Xiang Ai Ping An He Xie means *love, peace, and harmony.*

How have we created positive karma and negative karma in all our lifetimes? In ancient wisdom, karma is created through our shen kou yi. Shen means *body,* indicating our activities and behaviors. Kou means *mouth,* indicating our speech. Yi means *consciousness,* indicating our thoughts. Karma is created through our daily activities, behaviors, speech, and thoughts.

In Tao Science, positive karma is positive information, which is order and connection. If we offer positive service, including love, forgiveness, compassion, light, care, kindness, generosity, integrity, and much more, we are increasing positive information. We are literally creating positive karma. According to the Law of Karma, positive karma receives blessings for health, relationships, finances, intelligence, success, higher wisdom, higher realization for life, and reaching soul heart mind body enlightenment.

In Tao Science, negative karma is negative information, which is disconnection and disorder. If we offer negative service, including killing, harming, taking advantage of others, stealing, cheating, and much more, we are increasing negative information. We are literally creating negative karma. According to the Law of Karma, negative karma brings blockages and challenges in any aspect of our life, including health, relationships, finances, intelligence, failures, lack of wisdom, getting lost on the spiritual journey, and much more.

Forgiveness practice is one of the golden keys to self-clear negative karma. Negative karma is two-way traffic. On one hand, we have hurt others in our current lifetime and all our past lifetimes. On the other hand, others could have hurt us in all lifetimes. We also must consider ancestral karma. Many of us carry significant negative ancestral karma. Our ancestors have hurt others in all their lifetimes, and others have hurt our ancestors in all lifetimes.

Therefore, forgiveness practice is a two-way practice. The second sacred phrase for Da Kuan Shu is *I forgive you.* It does not matter how others have hurt us or our ancestors, we offer forgiveness to them unconditionally.

The third sacred phrase for Da Kuan Shu is *you forgive me.* This is not a command or a statement. It must be a sincere and humble request. When we and our ancestors have hurt others, we need to deeply realize our mistakes and sincerely apologize for the hurt and harm we have caused. Then, we can humbly ask those we have hurt to forgive us. *Sincerity and honesty move Heaven.* Expanding the wisdom of this ancient statement, *sincerity and honesty move souls.* The more we realize our mistakes, with great self-honesty and more, the more sincerely and humbly we can apologize and ask for forgiveness. The more Heaven and the souls we have hurt or harmed hear, see, feel, and know our true heart and mind, the more forgiveness they will grant us and bless us with.

The fourth sacred phrase for Da Kuan Shu is *bring love, peace, and harmony.* If forgiveness is given and forgiveness is received, then love, peace, and harmony among countless souls is created.

Understanding the wisdom of karma is very important. If someone hurts you in any way through anger, abuse, taking advantage, harming, or killing, understand that you could have hurt them in the same way in a past life. The Law of Karma tells us this very clearly. In Tao Science terms, we could have created negative information including anger, abuse, taking advantage, harming, or killing in a past life, which is negative quantum entanglement. The wisdom of Tao Science teaches us that information is divided into positive and negative. Therefore, quantum entanglement is also divided into positive quantum entanglement and negative quantum entanglement. Positive quantum entanglement is quantum entanglement that increases overall connection and order. Negative quantum entanglement is quantum entanglement that decreases overall connection and order. Negative karma is the negative entanglement created in past lives that is related to entanglement in this life.

The four sacred phrases for Da Kuan Shu are a sacred treasure to self-clear negative karma, which is negative information, in order to transform them to positive karma and positive information.

Here is a practical format for you to do forgiveness practice:

Apply Soul Power (say *hello*):

Dear Tao Source,

Dear Divine,

Dear Heaven, including all kinds of spiritual fathers and mothers,

Dear Mother Earth,

Dear all humanity,

Dear all the people, animals, nature, organizations, and more that I or my ancestors have hurt or harmed in any way in any lifetime,

I love, honor, and respect you all.

I deeply apologize for all our mistakes.

I ask for your forgiveness. I know in my heart and soul that to only ask for forgiveness is not enough. I have to serve more. To serve is to make others happier and healthier. I will serve you. I will serve wan ling.

For any soul who has ever hurt my ancestors or me, I totally and unconditionally forgive you.

Then chant (Sound Power):

I forgive you. You forgive me. Bring love, peace, and harmony.

I forgive you. You forgive me. Bring love, peace, and harmony.

I forgive you. You forgive me. Bring love, peace, and harmony.
I forgive you. You forgive me. Bring love, peace, and harmony. ...

Chant for at least five minutes. The longer and the more often you can chant, the better. There is no time limit. The more you do this practice, the faster and better you can transform your negative jing qi shen to positive jing qi shen. Transformation of every aspect of your life will follow.

Forgiveness practice is the top practice to self-clear negative karma. The transformation you can achieve is beyond comprehension.

Da Fu Wu (*Greatest Service*)

Da means *greatest*. Fu Wu means *service*. Da Fu Wu means *greatest service*.

To serve is to make others happier and healthier. This kind of service is creating positive information, energy, and matter. In Tao Science terms, it creates positive quantum entanglement. This positive quantum entanglement will entangle you to the person you serve. This entanglement could be instant. The entanglement could also occur later—even in a future lifetime. In Tao Science, this wisdom is vital to explain how negative information can affect one's life. In the same way, positive information also affects one's life when the time is ready.

Let us practice again to clear our negative karma:

Soul Power. Say *hello*:

Dear Tao Source,
Dear Divine,

Dear Heaven, including all kinds of spiritual fathers and mothers,

Dear Mother Earth,

Dear all humanity,

Dear all the people, animals, nature, organizations, and more that I or my ancestors have hurt or harmed in any way in any lifetime,

I love, honor, and respect you.

I deeply apologize for all our mistakes.

I ask for your forgiveness. I know in my heart and soul that to only ask for forgiveness is not enough. I have to serve more. To serve is to make others happier and healthier. I will serve you. I will serve wan ling.

For any soul who has ever hurt my ancestors or me, I totally and unconditionally forgive you.

Sound Power. Chant:

I forgive you. You forgive me. Bring love, peace, and harmony.

I forgive you. You forgive me. Bring love, peace, and harmony.

I forgive you. You forgive me. Bring love, peace, and harmony.

I forgive you. You forgive me. Bring love, peace, and harmony. …

Forgiveness practice is daily practice to transform negative karma, which is negative information. We can never do forgiveness practice enough. Therefore, we emphasize forgiveness practice again. Now let us chant *Da Fu Wu* (greatest service) to increase our positive information.

Chant:

Da Fu Wu
Da Fu Wu
Da Fu Wu
Da Fu Wu ...

Greatest service
Greatest service
Greatest service
Greatest service ...

Chant for at least five minutes. The more you can chant, the better. There is no time limit. The longer you chant and the more often you chant, the better you can transform your life.

We also need to do service practice. We have to act to serve. We have to transform any negativity in our activities, behaviors, speech, and thoughts to positivity. We need to do all kinds of humanitarian service.

Forgiveness practice and service practice are vital to transform our negative information to positive information. Service practice is the highest spiritual practice. Service practice is the highest method and gate to return to Tao. In one sentence:

The purpose of life is to serve, which is Xing Shan Ji De (*do kind things, accumulate virtue*) through shen kou yi (*actions and behaviors, speech, thoughts*), in order to transform negative jing qi shen to positive jing qi shen in every aspect of life, and finally to become a saint or a buddha and to reach Tao.

We wish every reader, every human being on Mother Earth, and every soul in countless planets, stars, galaxies, and universes to do forgiveness practice and service practice as much as possible. The benefits for yourself, for humanity, for Mother Earth, and for all souls would be priceless.

Think of one area or a few areas of your life that you would like to improve. Find out what karmic lessons you can learn from your challenges. In your heart, deeply apologize for the mistakes your ancestors and you have made in this aspect. Sincerely ask for forgiveness. Learn the lessons and transform your thoughts, feelings, speech, and all actions. Serve others unconditionally. In this way, you can overcome your challenges quickly.

Universal Law of Universal Service

In April 2003, I (Master Sha) held a Power Healing workshop for about one hundred people at the Land of Medicine Buddha in Soquel, California. As I was teaching, the Divine appeared. I told the students, "The Divine is here. Could you give me a moment?" I knelt and bowed down to the floor to honor the Divine. (At age six, I was taught to bow down to my tai chi master. At age ten, I bowed down to my qi gong master. At age twelve, I bowed down to my kung fu master. Being Chinese, I learned this courtesy throughout my childhood.) I explained to the students, "Please understand that this is the way I honor the Divine, my spiritual fathers, and my spiritual mothers. Now I will have a conversation with the Divine."

I began by saying silently, "Dear Divine, I am very honored you are here."

The Divine, who was in front of me above my head, replied, "Zhi Gang, I come today to pass a spiritual law to you. This spiritual law is named the Universal Law of Universal Service.

It is one of the highest spiritual laws in the universe. It applies to the spiritual world and the physical world."

The Divine pointed to the Divine. "I am a universal servant." The Divine pointed to me. "You are a universal servant." The Divine swept his hand across the audience. "Everyone and everything is a universal servant. A universal servant offers universal service unconditionally. Universal service includes universal love, forgiveness, peace, healing, blessing, harmony, and enlightenment."

The Divine explained, "If one offers a little service, one receives little blessing from the universe and from me. If one offers more service, one receives more blessing. If one offers unconditional service, one receives unlimited blessing."

The Divine paused for a moment before continuing. "There is another kind of service, which is unpleasant service. Unpleasant service includes killing, harming, taking advantage of others, cheating, stealing, complaining, and more. If one offers a little unpleasant service, one learns little lessons from the universe and from me. If one offers more unpleasant service, one learns more lessons. If one offers huge unpleasant service, one learns huge lessons."

I asked, "What kinds of lessons could one learn?"

The Divine replied, "The lessons include sickness, accidents, injuries, financial challenges, broken relationships, emotional imbalances, mental confusion, and disorder." The Divine emphasized, "This is how the universe operates. This is one of my most important spiritual laws for all souls in the universe to follow."

I asked the Divine, "Dear Divine, is your universal law karma law?"

The Divine replied, "Exactly."

I immediately made a silent vow to the Divine:

> *Dear Divine,*
>
> *I am extremely honored to receive your Law of Universal Service. I am making a vow to you, to all humanity, and to all souls in all universes that I will be an unconditional universal servant. I will give my total GOLD (gratitude, obedience, loyalty, devotion) to you and to serving you. I am honored to be your servant and a servant of all humanity and all souls.*

I received the Universal Law of Universal Service, which is the Law of Karma. I made a vow to serve unconditionally. The Divine gave me the honor to offer the Divine's karma cleansing in July 2003. Tao gave me the honor to offer Tao's karma cleansing in 2008. For nearly eleven years, until the end of 2013, I offered the Divine and Tao's karma cleansing to humanity. There are so many heart-touching and moving stories from my Divine and Tao karma cleansing service. I have trained and empowered almost one hundred fifty advanced students to offer Divine and Tao services.

11

Law of Tao Yin Yang Creation

WHERE DO EVERYONE and everything come from? How are everyone and everything created? What are space and time? Some of you may have asked these questions since childhood.

Understanding the secrets of creation is some of the highest wisdom, knowledge, and enlightenment a human being could ever dream about. This wisdom and knowledge will not only empower us to become powerful manifestors; more important, it is essential for liberating us from illusion, suffering, limitation, lack, ignorance, and bondage. This wisdom is the doorway to a higher level of consciousness and enlightenment. It is the key to go beyond life and death and reach immortality. In this chapter, we are honored to release another vital law of Tao Science, the Law of Tao Yin Yang Creation. This universal law explains the fundamental secrets of creation.

The Law of Tao Yin Yang Creation involves two basic concepts: Tao and yin yang. Tao is emptiness. It is the Source. It is the Creator of everyone and everything. Yin yang is a duality pair. Everyone and everything created by Tao consists of yin and yang. In fact, Tao Normal Creation explains that the first thing Tao Oneness created is Two, which are Heaven and Earth, yang and yin. The Law of Tao Yin Yang Creation explains *how* everyone and everything are created. Tao Normal Creation and Tao Reverse Creation, which we discussed in chapter seven, reveal the creation process in the simplest, deepest, and all-inclusive manner.

We have presented a spiritual understanding of Tao and the Tao creation process. Can we explain and comprehend them scientifically and mathematically?

Mathematics is the universal language of the mind. When we can express something mathematically, then, at least in principle, our mind has come to a complete understanding about it. Our mind gains the ability to create and transform it. Physics is powerful because it uses mathematical formulas to express the world. Therefore, it has great ability to create and transform the world. Using mathematics to express the truth of creation is one of the highest dreams anyone could ever have.

The Law of Tao Yin Yang Creation helps us formulate the physics theory about how our universe is created. This requires a deep understanding of space and time, even deeper than Einstein's. The insight of the Law of Tao Yin Yang Creation helps us gain a more profound comprehension about space and time, and this in turn enables us to derive the physics theory about creation. Our results could awaken humanity at its deepest core.

Essence of Yin Yang

What is the cause of creation, change, birth, growth, development, and death? How does Tao create all possibilities, infinite information, infinite energy, and infinite matter? The Law of Tao Yin Yang Creation answers these questions.

Tao is emptiness. It is the Source and Creator of everyone and everything. Tao creates yin yang. The interaction of Tao, yin, and yang creates everyone and everything.

Everyone and everything can be divided into yin and yang components. Yin and yang are opposite, related, co-created, inseparable, and interchangeable. Yin and yang comprise everyone and everything.

According to the Law of Tao Yin Yang Creation, yin and yang represent two opposite aspects existing within everyone and everything. Yin represents the aspect that has the nature of water—cool, passive, dark, contracting, feminine, and descending. Yang represents the aspect that has the nature of fire—hot, active, bright, expanding, masculine, and ascending.

The relationship between yin and yang includes the following four important characteristics:

Yin and yang are opposite and relative

Yin and yang are dual and opposing. However, they are relative, not absolute, because what is yin relative to one thing can be yang relative to another thing, and vice versa. For example, movement and stillness are a yin-yang pair. They are opposite but also relative. A car may appear still to you, but for another person who is moving away, the car may appear to be moving. Between the soul and the heart, the soul is yang and the heart is yin, because the soul transmits information (an active function) and the heart receives it (a passive function). However, between the heart and the mind, the heart is yang and the mind is yin. The heart passes the information to the mind, which receives it.

Yin and yang are inseparable and co-dependent

Yin cannot exist without yang. Yang cannot exist without yin. The existence of one depends on the other.

Yin and yang are co-created

When yin is created, yang is simultaneously and automatically created. When yang is created, yin is simultaneously and automatically created.

The interaction of yin and yang creates everyone and everything

The interaction of yin and yang is the cause of all creations and changes. Tao Normal Creation says this in the simplest way: *Two creates Three. Three creates all things.*

The foundational source book of traditional Chinese medicine, *Yellow Emperor's Internal Classic*, says: "Yin yang is the universal principle. It is the fundamental law followed by everyone and everything. It is the origin of all changes. It is the cause of birth and death. It is the reason of all creation."

Tao Yin Yang Creation of Everything

According to Tao wisdom, Tao is emptiness. Tao creates everyone and everything through the process of Tao Normal Creation:

Tao Sheng Yi	*Tao creates One*
Yi Sheng Er	*One creates Two*
Er Sheng San	*Two creates Three*
San Sheng Wan Wu	*Three creates all things*

Tao is emptiness. Tao is a state of Oneness (*Tao creates One*). When the time is right, Tao divides this state of Oneness into a yin-yang pair (*One creates Two*). This yin-yang pair has two states, yin and yang. Let's use the minus sign (-) to represent yin and the plus sign (+) to represent yang, and use an ordered pair to represent the two states of this yin-yang pair. Each element of this yin-yang pair can further subdivide and create four states: --, -+, +-, ++. Each of these four states can further subdivide and produce eight states: ---, --+, -+-, +--, -++, +-+,

++-, +++. This yin-yang subdivision process can continue indefinitely and generate infinite states and possibilities. In this way, everyone and everything is created. This Tao Yin Yang Creation process is illustrated in figure 12:

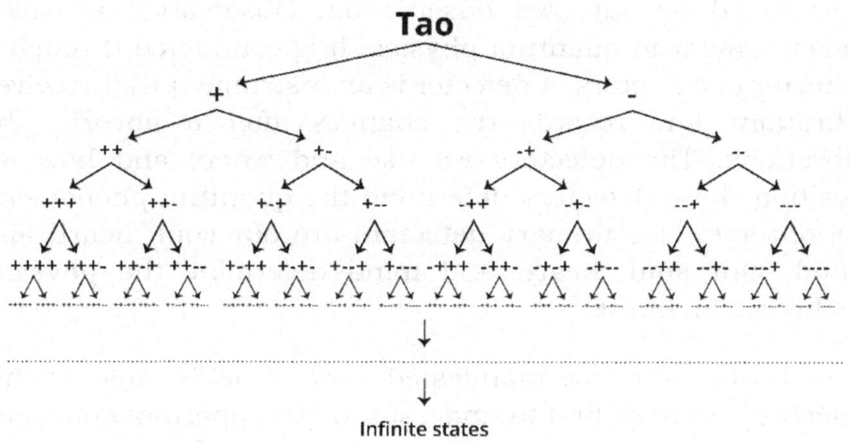

Figure 12. Tao Yin Yang Creation process

We can describe our world as a yin yang world because Tao yin-yang interaction creates everyone and everything. Now, let's examine how Tao yin-yang interaction creates everyone and everything via a mathematical formula. Can we understand scientifically and mathematically how physical reality—our universe—is created from emptiness?

Law of Tao Yin Yang Creation and quantum entanglement

The Law of Tao Yin Yang Creation tells us that everyone and everything consists of yin and yang, and that a yin-yang pair is one piece of information with two possible states: yin and yang. These two possible states contained in one yin-yang pair are connected. They are opposite, relative, co-created, inseparable, and co-dependent. The creation and transformation of one of them will instantly affect the other independent of space and time. Therefore, the Law of Tao Yin Yang Creation implies the existence of quantum entanglement phenomena.

How is the universe created?

Quantum physics tells us that our physical reality is manifested by our own observation. Observation is called measurement in quantum physics. It is conducted through a detector or detectors. A detector is an instrument that receives vibrations and records the changes after it absorbs the vibrations. The detectors we use and where and how we position these detectors determine the quantum phenomena we observe. Our primary detectors are our soul, heart, and mind. Our soul, heart, and mind determine the physical reality we manifest.

How is our universe manifested from Tao? To answer this question, we need first to understand two important concepts: space and time. As you will learn, the manifestation of our universe closely relates to the mysteries of space and time.

What Are Space and Time?

Space and time are the two mysteries that have been explored by all people in all cultures, traditions, ideologies, philosophies, and sciences in history. Space and time have many layers of meaning and applications.

In the seventeenth century, understanding about space and time emerged as a central issue in advancing science. There were two opposing theories about space and time. One believed that space and time do not exist. They are no more than the collection of, and an idealized abstraction from, the relations between objects and events.

The other theory believed that space and time are absolute in the sense that they exist permanently and independently of the existence of matter. Newtonian mechanics accepts the existence of absolute space and time as fundamental axioms.

In the eighteenth century, the German philosopher Immanuel Kant developed a theory of knowledge. Kant came to the realization that space and time are not objective features of the world. Rather, space and time are a framework used by us to organize our experiences.

In the twentieth century, Einstein found that space and time are relative and connected with each other. Space and time become two different aspects of one entity, spacetime. With the introduction of the gravitational field in his theory of general relativity, Einstein found that spacetime depends on matter, contradicting Newton's concept of absolute spacetime.

Quantum physics challenges both Newton's and Einstein's concepts about space and time in the most dramatic way. Quantum phenomena bluntly defy Einstein's assumption that the transfer of information cannot exceed the speed of light. Quantum entanglement phenomena provide a way to transfer information instantly.

Quantum physics also disputes our normal ideas about space and time in a profound way. For instance, quantum phenomena reveal that we cannot measure time and energy simultaneously with complete accuracy.

In our recent research, we discovered that there is an uncertainty relation regarding space and time. This uncertainty relation indicates that we cannot measure space and time simultaneously with complete accuracy. It takes space to measure time and it takes time to measure space.

A metaphysical understanding of quantum spacetime is still missing from quantum physics. In Tao Science, to create the

Grand Unification Theory—the theory of everything—it is critical and essential to realize the deepest meaning and the most profound application about space and time.

In Tao Science, measurement of space and time is not a trivial action. It is not simply assigning a number to a ruler, a clock, or an event. As we have shown in chapter three, the observation and measurement process is a determining factor for what is being observed. It is in fact part of the process that creates the phenomena we perceive. Quantum physics tells us that the phenomena we observe are in fact manifested by our own actions. This is a revolutionary revelation. Most physicists and scientists have not paid enough attention to the implications of this revelation. It is partly because it is too radical for them to accept and partly because full comprehension of this profound revelation calls for a higher level of consciousness. It took the Buddha many lifetimes of dedicated pursuit of the truth to reach the profound realization that physical reality is determined by one's actions. This insight is what led the Buddha to his final enlightenment. It is the key for understanding the cause of all suffering and finding the way to release it. As we will show, the wisdom from Tao, Buddha, and quantum physics plays a significant role in finding the deeper meaning and function of space and time.

Dr. Rulin Xiu's Journey to Understand Spacetime

Since the second year of college, I have spent many hours, days, months, and years contemplating space and time. Philosophically, I agreed with Kant's insight that space and time are a framework that we use to organize our experience. However, this understanding was not enough to help me create the Grand Unification Theory I was searching for.

When Master Sha and I worked on the Soul Mind Body Science System as the Grand Unification Theory, I started thinking intensively again about the meaning of space and time, but I still could not obtain a satisfactory insight.

At the time, I was teaching small classes at my home. One day, one of my students brought another man to my class. This man laid a book on the desk on my porch. It was a book about the Mayan calendar. He told me he thought this book would be useful for my research. Speaking further, I learned he is a composer. He had a great ability to receive extraordinary music and play it on a piano, even though he had never learned how to read music. It felt strange that this man, whom I had never met before, knew about my research. I had told hardly anyone, including my students, about my research. Besides, even with all the popular attention given to the end of the world prophecy linked to the Mayan calendar, I never before had any interest in learning about the Mayan calendar. I couldn't imagine how that book would ever be useful to me. I left the book on the desk without reading it.

A couple of days later, I was guided to ponder two mathematical formulas from string theory about space and time. In string theory, there are two types of space and time: the worldsheet and the observed space and time. The worldsheet has two dimensions. One dimension is space. The other dimension is time. The two-dimensional worldsheet is created by the movement of a one-dimensional string in time. The observed space and time is a projection from the worldsheet. It can be of higher dimensions, such as 10, 11, or 26 dimensions. The observed space and time is a function of the time and space on the worldsheet. The formulas I was looking at showed me that the observed space and time is made of all kinds of cycles.

Intrigued, I walked around my home contemplating the meaning of the formulas. Idly, I opened the book about the Mayan calendar lying on the table. To my astonishment, the page I opened to gave me a specific explanation about the two formulas I was trying to figure out.

To ancient Mayans, space and time are intrinsically cyclic. The Mayans had profound understanding and deep experiences about the cyclic nature of space and time. More important, they had a deep intuitive understanding about the two essential elements and driving forces that create our universe: Hunab Ku and Kuxan Suum. Hunab Ku is usually translated as "the one giver of movement and measure." Kuxan Suum literally means "the road to the sky leading to the umbilical cord of the universe." But I was still puzzled about how these two elements, space and time, create our universe. I also had a hard time understanding how space and time could be cyclic and what that really meant.

I became infatuated with the mysterious Mayan wisdom. I studied several more books related to the Mayan calendar and Mayan civilization. They helped me a little, but not enough to fully comprehend the deep knowledge and understanding ancient Mayans had about space and time.

One day, I was driving to meet with a good friend. On a road through a forest, ancient Mayans appeared to me in spiritual form. Their love and compassion filled my heart. It brought tears to my eyes. They told me that they had been helping me with my research. They invited me to visit their pyramids in Mexico. My heart was full of gratitude.

When I saw my friend a short while later, she gave me a CD. She told me that while she was waiting for me at a bookstore, she received a message to buy a CD of music for me. The title of the CD was "Mexico."

I immediately made travel plans. In four days, I was traveling through Mexico by bus. I visited eleven pyramids in ten days, immersing myself in feeling, experiencing, communicating with, and receiving from ancient Mayan saints at these pyramids. The power and wisdom I experienced and learned from them was truly incredible. At each pyramid, ancient Mayans gave me a special experience and downloaded their amazing wisdom about space and time to me. They showed me how to travel through space and time portals and how to transcend space and time to become immortal. What these ancient Mayans shared with me was amazing and immensely nourishing. It would take another book to relate what they shared with me. I am so grateful for the love, support, and contribution of ancient Mayans to Tao Science and humanity.

When I was home preparing for this trip, I received the idea to sleep on the pyramids, so I traveled light, with only a few clothes and a sleeping bag. I did buy a nice hammock when I arrived in Cancun. Most of the pyramids I visited were quite tourist-oriented. I needed to pay an entrance fee. It was not allowed to touch the pyramids, much less walk or sleep on them. However, there was one pyramid different from all the others. It is the largest pyramid I visited. It did not require an entrance fee. Most important, I could walk on this pyramid.

As soon as I got there, I realized this was the place the ancient Mayans had formally prepared for my visit. It was interesting that I actually came to this place by accident. After missing my bus to another destination, a local person told me about this place.

Although it was not spoken explicitly, I got the feeling that everyone in that park expected my visit. They welcomed me warmly with great hospitality. It really touched my heart. I had lunch and dinner in the courtyard of a beautiful home near the pyramid. The park was also nicely prepared for me, with only a couple of other people around. In the afternoon, a young man arrived and volunteered to show me around. He shared interesting information about the pyramid and the ancient Mayans.

The pyramid is spectacular. It has two hundred sixty steps and thirteen temple ruins. Each ruin represents one month of the Mayan calendar, which has thirteen months in a year and twenty days in a month. I walked up each step and visited each temple ruin, surveying the most beautiful scenery from the pyramid, feeling, smelling, tasting, and sensing what each day, each month, and each year meant for the ancient Mayans. Time became still and then went backwards. I immersed among and became one with the ancient Mayans.

In the evening, I walked around the entrance of the park trying to decide whether to set up my hammock on the pyramid. A woman came over to me. She seemed to know what I was thinking. She said there was a campground on the hill in the back and I could camp there. I took it as a sign that it was okay for me to put up my hammock on the pyramid.

I ended up sleeping on the pyramid that night. Lying comfortably on my hammock, tucked in my warm sleeping bag with my headlamp on, I wrote in my notebook the profound insights and formulas about space and time the ancient Mayans helped me understand.

What I learned from the ancient Mayans turned out to be exactly what Master Sha had been teaching us, which I did not realize enough to apply it to understanding space and time. The ancient Mayans helped me experience and realize that the space and time we observe are also made of jing, qi, and shen, like everyone and everything else. Jing is matter. Qi is energy. Shen includes soul, heart, and mind. Soul is the information. Heart is the receiver of the information. Mind is the processor of the information. Information is possibilities. Matter is the physical manifestation. Energy is the ability to do work. The jing, qi, and shen of space and time are carried by the vibrational field. The vibrational field consists of various vibrations with multitudes of frequencies and wavelengths. These various vibrations are the different cycles of space and time. This cyclic nature of space and time is the reincarnation behavior of space and time. Resonance is the way the heart receives information. Different spaces and different times carry their own different information, energy, and matter. They are also part of a larger cycle of space and time. Synchronicity of events in space and time is due to a profound soul connection. I am grateful to the ancient Mayans and other spiritual masters and beings who have helped me reach a deeper understanding of space and time.

The ancient Mayans also helped me understand that space and time are a yin-yang pair that is involved in creating the universe we observe. The specific formula to account for this process (and its connection to string theory) came to me during one of Master Sha's workshops. Master Sha's workshops are always filled with intense positive information and energy. Many breakthroughs and "aha!" moments have come to me during his workshops.

With Master Sha's blessing and the help of ancient Mayan saints and the Tao Science Committee, I realized that two yin-yang pairs are involved in creating the universe we observe. With this understanding, it is quite easy to write down the formula expressing our universe. I was so astonished at the simplicity, beauty, and power of this formula. It demonstrates so beautifully the profound wisdom and power of Tao Normal Creation and the Buddha's realization.

As with everything else, our universe is manifested from Tao through yin-yang interaction. We find that our universe is manifested through the yin-yang interaction of space and time.

To achieve complete comprehension of quantum physics and to create the Grand Unification Theory, the "theory of everything," it is critical and essential to realize the deepest meaning and the most profound significance of space and time. In Tao Science, we find that:

In their deepest meaning, space and time are the fundamental yin-yang pair that manifests our universe.

Time relates to movement and change. It is the active yang aspect of the yin-yang pair. Space relates to stillness and stability. It is the passive yin aspect of the yin-yang pair.

Space and time relate to two basic human actions. In quantum physics, our action is also called measurement. The world we observe is manifested by our measurement. Fundamentally, time and space are two types of measurement. Time is the measurement of change. For example, the movement of sand in an hourglass, the burning of incense, and the movements of the sun and moon have all been used as measurements of time. The duration of a day is based on measurement of the rotation of the Earth around its axis. The length of a lunar month is related to the movement of the moon around the Earth. A year is the measurement of a revolution of Earth around the sun.

Space is the measurement of invariability and stillness. The length, height, and width of an object are the measurement of its invariability and stillness.

As a yin-yang pair, space and time are opposite, co-created, relative, interdependent, inseparable, and interchangeable. Change and stillness are opposites. They are relative because something may appear to change to you but to be still to others. When you are a passenger in a moving train, the landscape can appear to you to be moving, but it is not moving from the perspective of someone standing beside the tracks. Therefore, change and stillness or invariability are interchangeable. Time and space are co-created because whenever we measure change we must refer to something invariable. Whenever we measure invariability, we must compare it to something changing. Change and stillness always come together. They are inseparable.

There is another fundamental measurement involved in manifesting our universe: inclusive and exclusive action. Our actions can include something or exclude something. The yin-yang pair of space and time can further divide into four states through inclusion and exclusion: yin space, yang space, yin time, and yang time. Yin space is the inclusive space. Yang space is the exclusive space. Yin time is the inclusive time. Yang time is the exclusive time. See figure 13.

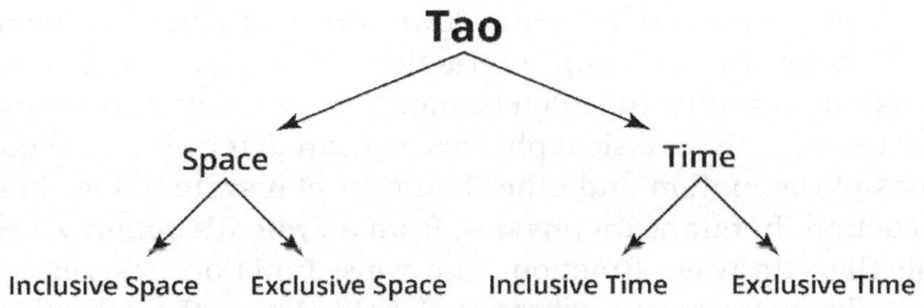

Figure 13. Tao creation of spacetime

Let's examine our actions more closely. We find that all measurement is based on spacetime measurement and inclusive-exclusive measurement. The measurement of velocity, acceleration, energy, momentum, temperature, spin, electricity, magnetic field, mass, charge, force, and more are all variations of spacetime measurement and inclusive-exclusive measurement. For example, to measure mass with a balance, we put the matter to be measured on one side of the balance and matter with known mass on the other side. When both sides of the balance are completely balanced, we know the two sides have completely equal mass. In this way, the unknown mass is measured. In this measuring process, spacetime measurement is used to make sure that the two arms of the balance are equal and still. Inclusive-exclusive measurement is applied when we add known masses to one side of the balance.

In one sentence, all measurements are variations and combinations of spacetime measurement and inclusive-exclusive measurement. We conclude that these two measurements are the fundamental yin-yang interactions that manifest our universe.

Manifestation of the Universe by Yin-Yang Interaction

To see how yin-yang interaction creates our universe mathematically, we need to write down the mathematical action created by the yin-yang interaction. In physics, action is a dynamic quantity that determines the property of a system. For example, in classical physics, we can determine the equations of the motion and other behavior of a system if we know its action. In quantum physics, from a system's action we can calculate its wave function. The wave function describes all the vibrations in the vibrational field. From the vibrational field, we can know the information, energy, and matter in a system.

The simplest action created by the interaction of the spacetime yin-yang pair is the product of space and time. It can be described simply as:

Action = Time x Space

If we consider a string to be a one-dimensional space, this simple action describes a string moving in time. For those who are familiar with string theory, this action is the action that underlies string theory. As we mentioned in chapter two, string theory studies the quantum dynamics of a string. In string theory, the vibrations of a string create forces, particles, and more. It is like a violin. The strings of the violin produce musical notes, which correspond to elementary particles, forces, and more in string theory. String theory has the potential to unify all fundamental forces and elementary particles. It is the most promising candidate to produce the Grand Unification Theory.

When we write down the simplest action that includes the interaction of both the space-time and inclusive-exclusive measurements, we find that this action is the same as the action that creates superstring theory. Superstring theory is the extension of string theory with supersymmetry. Supersymmetry is the symmetry that connects and unifies different kinds of fundamental particles.

Superstring theory is also called M-theory. Edward Witten and other brilliant string theorists have shown that all physics theories are included in M-theory. For example, general relativity can be derived from M-theory. All fundamental forces and elementary particles are also contained in superstring theory. M-theory includes all current physics theories.

M-theory has great potential to be the grand unified theory that can explain everyone and everything. However, it has difficulty in making many testable predictions. Something is still missing in current M-theory.

Tao Science includes string theory and M-theory, but it tells us a lot more about our universe than string theory and M-theory can. With the action created from the yin-yang interaction obtained above, we can write down the wave function created by this action. To calculate the wave function created by this action is to add all the possible states created by the action. We find this wave function could be the wave function of our universe because it tells us all the possible information, energy, and matter, as well as the large-scale structure of the universe. It also tells us how our universe is created, how it evolves, and what its destiny is.

For readers who want to explore this mathematical derivation, please refer to our research papers in the bibliography on page 251.

According to ancient spiritual wisdom, Tao creates Heaven and Earth. The interaction of Heaven and Earth creates everyone and everything, i.e., our universe. If we regard time as Heaven and Space as Earth, Tao Science confirms this ancient spiritual wisdom mathematically.

One of the greatest achievements of Tao Science is its ability to write the wave function of the universe from fundamental principles. Although it is rather simple to write the formula for the wave function of the universe, it is another matter to calculate it. However, even without completely calculating this formula, we can derive some amazing results from it. In our research, we have shown some intriguing results derived from the wave function of our universe.

Now we'll share with you what we have found about how our universe is created, the driving force for the expansion of our universe, the source of dark energy and dark matter, and other properties of our universe.

Our derivation of the wave function of our universe demonstrates the origin of our universe:

The yin-yang interaction of space and time manifests our universe from Tao, which is emptiness.

Tao creates and includes all possibilities. Our yin-yang action determines which possibility or possibilities are manifested from Tao.

Are there more universes?

Is our observed universe the only existing universe? Are there more universes? Our formula of the wave function of our universe exposes mathematically countless possible universes existing simultaneously. Our universe actually contains many possibilities. It exists simultaneously in all these possibilities. In this sense, we exist in multiple universes.

How is our specific universe manifested? Our specific universe is manifested by our actions. Each one of us is helping to shape our universe. This is how important and powerful we are. We are influencing the whole universe and many universes!

Multiple universes have been hypothesized in cosmology, astrophysics, religion, philosophy, psychology, and literature, especially in science fiction and fantasy. They are also called parallel universes, multiverse, meta-universe, alternate universes, quantum universes, parallel worlds, alternate realities, alternate timelines, interpenetrating dimensions, or dimensional planes. Physicists propose the multiverse to help tackle some unsolved problems in cosmology, including the cosmological constant problem. However, their proposition of the multiverse is arbitrary. It is ad hoc, so to speak. In Tao Science, we can derive the mathematical formula for the multiverse from the Law of Tao Yin Yang Creation.

The expanding universe, dark energy, and dark matter

Observations in astrophysics indicate that our universe is not only expanding, this expansion is also accelerating. This tells us that there must be an energy source driving this expansion.

What is this energy source? At this time, it is a complete mystery to physicists. Although physicists cannot identify or explain the source of this energy, they have named it *dark energy*. They can estimate how much dark energy exists from the acceleration of the universe's expansion. It turns out that more than two-thirds, roughly sixty-eight percent, of the universe is dark energy.

Furthermore, scientists found that they needed to introduce the concept of *dark matter* to explain the large-scale structures in the universe. By dark matter, scientists mean matter with mass that cannot be explained by any of the matter that is known, such as electrons, photons, atoms, molecules, planets, stars, galaxies, interstellar medium, black holes, white holes, antimatter, intergalactic dust, and more. They have extrapolated that dark matter makes up about twenty-seven percent of our universe.

It is estimated that less than five percent of the universe is made of the matter and energy we know of.

What could this "dark energy" be? The simplest possible source for dark energy is vacuum energy. In cosmology, it is called the cosmological constant. Einstein first introduced this term in one of his equations. Cosmological constant is a term appearing in Einstein's equation that comes from the energy of the vacuum.

As we have mentioned in chapter seven, in quantum physics, emptiness or a vacuum is not nothingness. Within the emptiness, vibrations come and go. Within the emptiness, there are information, energy, and matter. However, when one uses current quantum theory to calculate this energy, it turns out to be 10^{120} times larger than the current measurement of the total energy of the universe, including dark matter and dark energy. This discrepancy has been called "the worst theoretical prediction in the history of physics." This is the famous cosmological constant problem in physics.

From the wave function of our universe that we have derived, we find there exist vibrations that are very fine. They are also very dark in the sense that it is very difficult to detect them. In fact, it would take the lifetime of the entire universe or a detector as large as our universe to detect some of these vibrations. The existence of these vibrations in the wave function of the universe explains the existence of dark energy and dark matter.

From the wave function of our universe, we can estimate the vacuum energy of our universe. Our calculation[11] agrees with current experimental data about dark energy.

Our research demonstrates to us that Tao, true emptiness, is the Source. The energy our universe receives from Tao is determined by the yin-yang actions. This energy is rather limited. Nevertheless, it is enough to propel the continued and accelerated expansion of our universe.

---

[11] Dr. Zhi Gang Sha and Dr. Rulin Xiu. "Dark Energy and Estimate of Cosmological Constant from String Theory" to appear in *Journal of Astrophysics & Aerospace Technology* (accepted on March 27, 2017).

Our Universe Is a Holographic Projection

There are two kinds of spacetime. One is the space time yin-yang pair that comes from yin-yang action as we explained earlier in this chapter. They are the yin yang elements that manifest our universe. We can call this spacetime *internal spacetime.* In string theory, internal spacetime is called worldsheet.

The other spacetime is what we observe in our daily lives, such as the spacetime indicated by our clocks and the distance between our homes and offices. Let's call this spacetime *external spacetime.* From the wave function of the universe, we can see that external spacetime and all the observed particles and forces are a projection from internal spacetime.

Internal spacetime has an interesting property. If we stretch or compress it, it does not affect the phenomena we observe in external spacetime. In other words, internal spacetime is a hologram. This hologram contains all the information about our universe. Our universe is a projection from this hologram.

The wisdom that our universe is a hologram has been known for millennia in many cultures and traditions. It is wonderful to see that this ancient wisdom can be explained and expressed scientifically and mathematically in Tao Science.

The fact that our observed space and time is a projection from a hologram has great consequences. For example, string theorists have discovered a beautiful mathematical result. They found that the classical equations of motion as well as Einstein's general relativity can be derived from string theory as a result of our universe being a projection from a hologram.

In Tao Science, our universe consists of many possibilities. In the myriad of the multiverse, our world changes constantly and quickly depending on our feelings, thoughts, and actions. For example, when we are happy and joyful, we are connecting with and manifesting a joyful field in the universe. In quantum science, we understand quantum entanglement is a quality of quantum fields. In Tao Science, we have discovered that quantum entanglement can be divided between positive quantum entanglement and negative quantum entanglement. A joyful field is positive quantum entanglement. Positive quantum entanglement carries positive information, energy, and matter.

If we are upset, we connect with the "upset field" in the universe, which is negative quantum entanglement carrying negative information, energy, and matter.

On the other hand, in Tao Science, certain things do not change even when we are in a different mood, have different thoughts, are in different locations, and take different actions because of the fact that our universe is a projection from the hologram. The unchanging quantities make up the laws and the fundamental particles and forces we observe.

Space and Time Cycles and Reincarnation

The cycle of birth, growth, aging, and death is a natural law in the yin yang world.

Everyone and everything in our universe is a vibrational field made of various vibrations. Each vibration vibrates at its own frequency and wavelength. These vibrations constitute vastly diverse time cycles (time period or frequency) and space cycles (wavelength) in our observed space and time. The time cycles are expressed by periods and frequencies. A period is the time it takes to complete one oscillation. Frequency expresses how fast a cycle oscillates. Wavelength measures the size of a cycle. Wavelength is the distance between the peaks of two adjacent oscillations.

Countless space and time cycles exist in our universe. Some are too small to be observed. Some are too large to be noticed. The vibrations that have high frequencies and short wavelengths are small time cycles and space cycles. They make up the microscopic world. They are the phenomena associated with light, electrons, quarks, and more. The vibrations that have very long time cycles and space cycles make up the macroscopic world, including planets, solar systems, stars, galaxies, and universes.

Space and time are essentially cyclic. We experience the cycles of space and time in every moment. Different colors of light have different cycles of space and time. For example, the color red typically has frequencies between 400 and 484 THz (THz stands for terahertz, which is a frequency of 10^{12} per second). Its wavelength is between 620 and 750 nm (nm is one nanometer, which is 10^{-9} meter). Green light has frequencies between 526 and 606 THz and wavelength between 495 and 570 nm. X-rays have a wavelength ranging from 0.01 to 10 nanometers, corresponding to frequencies in the range 30 petahertz (10^{16} Hz) to 30 exahertz (10^{19} Hz). Infrared light has a wavelength ranging from 700 nanometers (frequency 430 THz) to 1,000,000 nanometers (frequency 300 GHz).

Earth's rotation around its axis creates our cycle of days. The moon's revolution around Earth forms the cycle of lunar months. Earth's revolutions around the sun create the cycle of years. The solar system orbits around the center of the Milky Way galaxy. This produces the galactic year, also called cosmic year. A cosmic year lasts 225 million terrestrial years.

Our world is made up of various space and time cycles. The space cycle creates repetitive patterns in nature. Time cycles are time reincarnation. A day reincarnates, day after day. A month reincarnates, month after month. A year reincarnates, year after year.

Time reincarnation is a fundamental natural phenomenon. Everything reincarnates on different levels. Light reincarnates. Atoms reincarnate. Human beings reincarnate. Mother Earth reincarnates. Heaven reincarnates. Universes reincarnate.

A human being contains countless different vibrations with various time cycles. The reincarnation of a human being is much more complicated than the minute hand of a watch rotating again and again every hour. The reincarnation of a human being is determined by all the information, energy, and matter in one's vibrational field. It is determined by one's karma. Karma is the service one's ancestors and oneself have conducted in the past.

Time reincarnation is a natural and universal phenomenon in the yin yang world.

Ancient Mayans understood the cycles of space and time, and the phenomenon of time reincarnation. They were very aware of the large heavenly cycles regarding time and space. They knew how these heavenly cycles affect each human being and even the history of humanity. The Mayan calendar describes some of these time cycles. The Mayan calendar reveals to us not only how some of these cycles affect our civilization, our society, and our lives, but also how we can use these cycles for our well-being. Learning and using these different cycles is great wisdom. It can lead to great power for creation and manifestation.

I Ching and Tao Science

The *I Ching* or *Classic of Changes* is the oldest extant Chinese classical text. It reveals the universal program existing within our universe, and in everyone and everything. The Ba Gua (*eight symbols*) contains the fundamental building blocks for all possible changes in the universe.

The German scientist, Dr. Martin Schönberger, discovered an astonishing similarity between the genetic code of life and *I Ching*. In his book, *The I Ching and the Genetic Code: The Hidden Key to Life*, Dr. Schönberger points out that life's genetic code is stored in a yin-yang pair, the double helix of DNA. Four letters are used to label the double helix. Three of these letters at a time form a code word for protein synthesis. To date, sixty-four of these triplets have had their properties and informative "power" explored. One or more triplets program the structure of one of the twenty-two possible amino acids. Specific sequences of such triplets program the form and structure of all living creatures.

Can we understand how *I Ching* works in a scientific way? We were fascinated to notice that the reason that all the changes in the universe can be described by *I Ching*, Ba Gua, and the sixty-four different situations is precisely because our universe is created by the interaction of two yin-yang pairs. The two yin-yang pairs have four elements. Since all the yin-yang interaction occurs through an element splitting into another yin-yang pair, there are in total 4 x 4 x 4 = 64 different ways of changes. These sixty-four ways are all the possible changes in our universe. This is why the Ba Gua in *I Ching* includes all the changes in the world.

Let us compare Tao Science and *I Ching* in greater detail. In Tao Science, our universe is created through the interaction of two yin-yang pairs from Tao. In *I Ching*, the universe and everyone and everything are created through the following process:

Tai Ji Sheng Liang Yi
Tao creates two sides.

Liang Yi Sheng Si Xiang
The two sides create four images.

Si Xiang Sheng Ba Gua
Four images create Ba Gua, the eight symbols that generate sixty-four situations.

Ba Gua Ding Ji Xiong
The Ba Gua determine whether the situation is beneficial or dangerous.

Ji Xiong Cheng Da Ye
The beneficial and dangerous situation fulfills the big mission.

It is interesting to see that the creation process described above by *I Ching* is equivalent to the Tao yin-yang creation of our universe.

First, one yin-yang pair, the space and time yin-yang pair, is created. This is liang yi, the two sides. (Liang means *two.* Yi means *side.*) Then, space and time each create another yin-yang pair, the inclusion and exclusion yin-yang pair. (See figure 12 on page 207.) Now, four images have been created. These four images are the si xiang. (Si means *four.* Xiang means *image.*) From these four images, Ba Gua and a total of sixty-four different ways of changes (the sixty-four situations) are derived. These sixty-four different changes determine what happens in the universe.

I Ching has been revered as the foremost classic of divination, cosmology, and philosophy for more than five thousand years in China. It reveals the profound wisdom that although there are countless possibilities existing in our world and our universe, there are only sixty-four different ways to change. These sixty-four ways to change determine what happens in our lives and our world. *I Ching* provides a simple and effective way to study all the possible changes in our universe and our lives. It will have significant application in cosmology, particle physics, and other areas of physics and science, such as medicine, economics, finance, environmental protection, and future technologies.

Many cultures, traditions, philosophies, and ideologies have realized the profound truth about the four images that build our world and our existence. For instance, it is interesting to notice the correspondence among the si xiang (*four images*) in *I Ching*, the two yin-yang pairs in Tao Science, the worldsheet of superstring theory, the cross in Christianity, and the concept of world tree in the mythologies and folklore of Northern Asia and Siberia, as well as in the Mayan, Aztec, Izapan, Mixtec, Olmec, and other Mesoamerican cultures and indigenous cultures of the Americas. For example, in the mythology of the Samoyeds, the world tree connects different realities (underworld, this world, upper world) together. In Mayan culture, world trees embodied the four cardinal directions, which represented also the fourfold nature of a central world tree, a symbolic *axis mundi* connecting the planes of the underworld and the sky with that of the terrestrial world. Different traditions use various words to express the same truth. Tao Science provides a scientific and mathematical way to understand, describe, and unify all of this ancient wisdom.

Five Elements Theory and Tao Science

The Five Elements Theory plays a crucial role in traditional Chinese medicine and every aspect of life in China. This wisdom is also known in other countries, religions, belief systems, cultures, and disciplines. The wisdom of Five Elements tells us that everyone and everything are made of five basic elements: wood, fire, earth, metal, and water. Each of the Five Elements is itself a yin-yang pair. For example, the wood element contains the yin wood element and the yang wood element.

Can this wisdom be explained scientifically? We can answer *yes*. According to Tao Science, our universe is created through the interaction of two yin-yang pairs, which are the "four images." These four images have a total of sixty-four possible ways to change. These sixty-four ways can be expressed with five bits of information ($2^5 = 32$) plus a yin-yang pair ($32 \times 2 = 64$). These five pieces of information are the Five Elements. Each of the Five Elements has yin and yang elements. From this, we can see that Five Elements Theory is a simple and effective way to understand the different changes existing in our universe, our world, and our lives.

Five Elements Theory tells us that although our bodies, our lives, our society, our world, and our universe can be very complicated, we need to be concerned with only Five Elements. As long as we can balance the Five Elements and ensure that they function well, everything will be fine. To heal and transform anything, we only need to focus on these Five Elements. What a simple and profoundly wise way to transform our health, longevity, every aspect of our lives, and the world.

So much new research is published that it is almost impossible to keep abreast of all significant developments and grasp the overall movement and directions of scientific progress. Ancient wisdom sees the bigger picture and looks at things from a broader and deeper perspective. It grasps the essence and critical points of everyone and everything.

With this scientific understanding of Five Elements Theory, this ancient wisdom can have greater application in health, medicine, economies, physics, biology, politics, environmental protection, world peace, relationships, and more in the future worldwide.

Application of the Law of Tao Yin Yang Creation in Daily Life

How can we apply the Law of Tao Yin Yang Creation to benefit our lives? Let us give you some examples.

Easy and difficult are a yin-yang pair. Easy and difficult are opposites. They are also relative because something can be easy for you but difficult for others. They are co-created because whenever you think something is easy, you are comparing it with something you consider difficult. Easy and difficult are inseparable because something easy cannot exist without comparing it to something difficult. The existence of easy or difficult things depends on each other. They cannot be separated. Easy and difficult are interchangeable because something easy can become difficult and something difficult can become easy. Easy and difficult may appear to be different. However, because they are co-created, inseparable, relative, and interchangeable, they are just different aspects of one thing. They are one.

Within everyone, everything, and every situation, there are two aspects, yin and yang. Let's continue to study the easy-difficult yin-yang pair to gain deeper understanding. Nothing is only easy. Nothing is only difficult. There are always easy and difficult aspects existing simultaneously in everyone, everything, and every situation. When you think you are creating something easy, you are simultaneously creating the difficult aspect. The easy and the difficult come hand in hand. If you see only the easy aspect without paying attention to the difficult side, it will cause you problems. If you only recognize the difficulty and do not think about coming up with an easy solution, you will also be stuck.

Only knowing one side and not the two sides of everyone and everything is the cause of most problems and challenges in our lives. Some people only want good things in their lives, they refuse to accept the negative aspect. This attitude is against yin yang law. It can make them frustrated, become unhealthy, and even lead to death.

For instance, for good health, people know the benefits of having good nutrition. But they do not realize that too much nutrition can be harmful. Nutritional imbalances are a major cause of many sicknesses. Currently, over-nutrition is a huge issue causing sickness on Mother Earth. Many people are taking too many—and too much—nutrients. Their bodies cannot digest and absorb these nutrients properly. They lose yin-yang balance.

Eating fruit is good for health, but if someone eats only fruit, becoming a fruitarian, he could become sick.

Some people believe physical exercise is good for health. They may forget that too much exercise could be harmful. This is the reason that some athletes live short lives.

Some people believe meditation is beneficial. They sit and meditate seriously, but they neglect movement. This imbalance can cause harm to their physical, emotional, mental, and spiritual health.

According to traditional Chinese medicine, when one is too yin or too yang, one is not far from death. Insisting on one thing, one way, one aspect, one absolute principle could cause harm to one's health and even threaten one's life.

Knowing that there are always two sides, yin and yang, to everyone and everything and that yin and yang should be balanced, will help solve all problems in our lives, our societies, and our world. The way to health and longevity is to see both the yin and yang aspects within everything and to balance yin and yang.

We should balance our nutrition. We should balance movement and stillness. We should balance exercise and rest. We should balance work and relaxation. We should balance solitude and socializing. We should balance indoor and outdoor activities. We should balance feminine and masculine energy. We should balance giving and receiving. We should balance every aspect of our lives.

You can apply the Law of Tao Yin Yang Creation to raise your children. In this case, love and discipline are the yin-yang pair. Both love and discipline play important roles in helping children grow. Balancing love and discipline is crucial for the development of your children. The Law of Tao Yin Yang Creation teaches us how to love and discipline children. If you want to discipline your children, do it with love. When you give love to your children, remember to love your children with discipline. In this way, you can help your children grow in a healthier way.

It is important to apply the Law of Tao Yin Yang Creation to environmental protection and economic development. People and countries seek economic growth, but they may forget loving, honoring, and respecting the environment. This has caused damage and disasters to the environment and to people. On the other hand, if we only think about protecting the environment without thinking about how to help people improve their lives, this can also have negative effects. The Law of Tao Yin Yang Creation teaches us that when we develop the economy, we should make sure that we do it in a way that is beneficial to the environment. When we protect the environment, we need to protect it in a way that helps grow the economy. When we do this, prosperity and harmony can come to humanity and Mother Earth. You may ask whether it is possible to have both economic growth and environmental protection. The answer is *yes*. There are many ways to do this. Why can't we do it right now? It is not due to the lack of a solution but rather, due to ignorance and greed. When we follow the Law of Tao Yin Yang Creation and take wise action, it is possible to protect the environment *and* grow the economy.

In summary, we live in the yin yang world. The Law of Tao Yin Yang Creation tells us:

- Everyone and everything has both yin and yang aspects. Yin and yang are opposite, relative, co-created, inseparable, and interchangeable. Yin and yang appears to be two distinct things, but they are only aspects of one thing.

- Everyone and everything in existence is created through yin-yang interaction.

- Yin yang is the basic universal element and driving force behind all creation and change in the universe.

To have success, peace, and harmony, it is important to apply the Law of Tao Yin Yang Creation in our lives by doing the following:

- Recognize both the yin and yang aspects in everyone, everything, and every situation.

- Love, honor, and appreciate both yin and yang.

- Balance yin and yang.

Apply the Law of Tao Yin Yang Creation to the challenges and difficulties in your life. Identify the yin and yang aspects of your issue. Start to pay attention to both aspects. Find the way to balance the two aspects. When you do this, you are on the way to resolve your difficulties and bring love, peace, and harmony into your life.

Is Our World an Illusion?

We manifest our universe. Our own action of measuring space and time as well as action of inclusion and exclusion manifest our universe. If we change our action, the universe we observe will change accordingly. In this sense, the world is an illusion. It has great flexibility rather than true solidity. When Buddha gained enlightenment while sitting under a bodhi tree, he realized that physical reality is manifested by our own action. Our action is the source of our physical reality. Buddha taught people to see through the illusion and not attach to the physical reality.

On the other hand, our world does truly exist in the sense that if you set up detectors to detect the vibrations of this world, you will find those vibrations. These vibrations do exist. Our physical reality does truly exist in this sense. In fact, our physical reality is important for our existence. Through the physical realm, we learn our karmic lessons and have the opportunity to advance our souls, hearts, minds, and bodies to higher levels. Physical reality is essential not only for our physical existence but also for our spiritual growth. However, none of our physical reality, even the spiritual images we may see, is Tao, Buddha, or our true selves. They are not the ultimate truth. We should not attach to them.

Within the emptiness of Tao, all possible vibrations exist. The existence of these vibrations does not depend on our own actions. However, the vibrations within the emptiness that are observed or manifested do depend on our actions. The reality we experience depends on our souls, hearts, and minds. Therefore, just like everything else, our reality has two aspects. It is an illusion as well as a true existence. To see both sides of our reality will help us reach higher levels of enlightenment and freedom. When we can remain unattached to the physical realm and use our physical reality as a tool to purify, open, develop, and advance our souls, hearts, minds, and bodies, as well as to serve our spiritual journeys, we can uplift our physical existence, our souls, our hearts, and our minds back to the Source.

Beyond Yin and Yang

Using yin yang elements, we manifest our physical reality, the yin yang world. Balancing yin yang elements, we achieve health, prosperity, beauty, and peace. Going beyond yin yang, we go back to the Source, Tao.

Tao is the Source. Tao contains infinite information, energy, and matter.

The natural laws taught in science are the laws of the yin yang world. To go beyond yin and yang is to go back to Tao. To go back to Tao is to no longer be limited by these natural laws. We may transcend gravity and levitate. We may transcend the cycles of space and time and stop reincarnation.

Going beyond yin and yang, which is returning back to the Source, can open all possibilities for us. We become one with all powers, all wisdom, and all creation. Eternal life is possible. We will have the ultimate bliss and freedom in every moment. This is the state of ultimate enlightenment and immortality. To return to Tao and reach the state of ultimate enlightenment and immortality is the goal for all beings. Although it is very easy to say, it is far from easy to achieve. However, it is definitely possible for everyone. Tao Science is the science of enlightenment and immortality. We look forward to presenting the further development of the science of enlightenment and immortality in future books.

To go beyond yin and yang, it is critical and essential to understand the Law of Shen Qi Jing, the grand unification formula, the Law of Karma, and the Law of Tao Yin Yang Creation. We wish you, our beloved reader, and all beings to reach the highest fulfillment, the ultimate enlightenment, and immortality.

Conclusion

TAO SCIENCE is created now.

Physics is the foundation of natural science. We have reviewed the major milestones of classical physics, including Newtonian mechanics, thermodynamics, optics, and electromagnetism, and of modern physics, including Einstein's special and general relativity, quantum physics, astrophysics, particle physics, string theory, and the search for the theory of everything.

Tao Science is a breakthrough for science because:

- Tao Science explains in a scientific way the ancient sacred wisdom of shen qi jing. In Tao Science, we scientifically define shen as information. We show that shen includes soul, heart, and mind. Qi is energy. Jing is matter. Soul, heart, and mind are the three aspects of information. Soul is the content of information. Heart is the receiver of information. Mind is the processor of information. Tao Science defines energy as the actioner because it makes an action possible. Matter is physical existence. Matter is also the transformer. Every aspect of life is made of information, energy, and matter. The purpose of life is to enhance positive information and transform one's soul, heart, and mind. Physical reality, which is matter, is to help increase one's positive information to uplift one's soul, heart, and mind. Therefore, we can also define matter as the transformer.

- Tao Science shares Tao Normal Creation, which explains how the universe is formed, and Tao Reverse Creation, which explains how the universe develops and ends.

- Tao Science explains that Tao Oneness carries infinite highest, purest information, energy, and matter, as measured by negative entropy. This highest, purest information, energy, and matter can purify and transform all kinds of negative information, energy, and matter, as measured by entropy, in every aspect of life.

- Tao Science explains karma scientifically. Karma is divided into positive karma and negative karma. Positive karma is positive information, energy, and matter. Negative karma is negative information, energy, and matter.

- Why is there no solution or not enough of a solution for many sicknesses? Why are there many challenges in relationships, finances, and every aspect of our lives for which there is no complete solution?

  Tao Science explains that the root blockages for challenges in any aspect of life are negative karma, including negative information, energy, and matter. Transforming negative information, energy, and matter to positive information, energy, and matter is the most important transformation.

  These scientific findings and explanations can bring breakthrough transformation for science, medicine, health, relationships, finances, business, economies, politics, and every aspect of life.

We explained this wisdom in chapter ten on the Law of Karma.

- Tao Science explains ancient sacred wisdom that yin-yang interaction creates everyone and everything. We explained this wisdom in chapter eleven on the Law of Tao Yin Yang Creation.

- Tao Science creates and explains the Grand Unification Theory and practice. It presents the scientific formula of grand unification: S + E + M = 1. In chapter seven, we shared practical techniques to apply S + E + M = 1 to transform health, to rejuvenate, to prolong life, and to advance on the path to reach immortality.

In the messages and information we have received from Tao Source, the Tao Calligraphy Tao Chang contains dark energy and dark matter in countless planets, stars, galaxies, and universes, which is the "existence world" (You World in Mandarin Chinese). Modern science has not been able to explain this dark matter and dark energy.

We share an additional wisdom that has implications beyond comprehension: within a Tao Calligraphy Tao Chang are information, energy, and matter that are beyond the existence world. This realm is the Wu World or "emptiness world." In this Tao Oneness Wu World, there are Wu World dark energy and Wu World dark matter. We will present more of this wisdom in future books and articles.

Tao Science shares a new understanding of the universal laws. We wish the wisdom in this book will help develop many beneficial modern technologies and create many new ways to bring health and happiness to humanity and create a Love Peace Harmony World Family for humanity and Mother Earth, and a Love Peace Harmony Universal Family for countless planets, stars, galaxies, and universes.

I love my heart and soul
I love all humanity
Join hearts and souls together
Love, peace and harmony
Love, peace and harmony

Acknowledgments

WE THANK from the bottom of our hearts the beloved one hundred eleven saints of Heaven's Tao Science Committee and Heaven's Soul Mind Body Science System Committee who flowed this book through us. We flowed this entire book from them as they were, and are, above our heads. We are so honored to be their servants. We are so honored to be servants of humanity and all souls. We are eternally grateful.

We thank from the bottom of our hearts the Divine and Tao.

Master Sha thanks from the bottom of his heart all of his beloved spiritual fathers and mothers, including Dr. and Master Zhi Chen Guo. Dr. and Master Zhi Chen Guo was the founder of Body Space Medicine and Zhi Neng Medicine. He was one of the most powerful spiritual leaders, teachers, and healers in the world. He taught Master Sha the sacred wisdom, knowledge, and practical techniques of soul, mind, and body. Master Sha cannot honor and thank him enough.

Master Sha thanks from the bottom of his heart Professor Liu Da Jun, the world's leading *I Ching* and feng shui authority at Shandong University in China. Professor Liu has taught Master Sha profound secrets of *I Ching* and feng shui. Master Sha cannot honor and thank him enough.

Master Sha thanks from the bottom of his heart Dr. and Professor Liu De Hua. He is a medical doctor and was a university professor in China. He is the 372nd-generation lineage holder of the Chinese "Long Life Star," Peng Zu. Peng Zu was the teacher of Lao Zi, the author of *Dao De Jing*. Professor Liu De Hua has taught Master Sha the secrets, wisdom, knowledge, and practical techniques of longevity. Master Sha cannot honor and thank him enough.

We thank from the bottom of our hearts Lao Zi, Fu Xi, A Mi Tuo Fo, Shi Jia Mo Ni Fo, Ling Hui Sheng Shi, Da Shi Zhi, Babaji, Mayan saints, and many other saints and buddhas.

Master Sha also thanks his beloved sacred masters and teachers who wish to remain anonymous. They have taught him sacred wisdom of Xiu Lian (purification practice) and Tao. They are extremely humble and powerful. They have taught him priceless secrets, wisdom, knowledge, and practical techniques, but they do not want any recognition. Master Sha cannot honor and thank them enough.

We thank from the bottom of our hearts our physical fathers and mothers and all of our ancestors. We cannot honor our physical fathers and mothers enough. Their love, care, compassion, purity, generosity, kindness, integrity, confidence, and much more have influenced and touched our hearts and souls forever. We cannot thank them enough.

We thank from the bottom of our hearts our literary agent, William Gladstone, for his incredible contribution and selfless support. We cannot thank him enough.

We thank from the bottom of our hearts the chief editor, Master Allan Chuck, for his excellent editing of this book and almost all of Master Sha's other books. He is one of Master Sha's Worldwide Representatives. He has contributed greatly to the mission and his unconditional universal service is one of the greatest examples for all. We cannot thank him enough.

We thank from the bottom of our hearts Heaven's Library Publication Corp.'s senior editor, Master Elaine Ward, for her excellent editing of this book and most of Master Sha's other books. She is also one of Master Sha's Worldwide Representatives. We thank her deeply for her great contributions to the mission. We cannot thank her enough.

We thank from the bottom of our hearts Master Lynda Chaplin, another one of Master Sha's Worldwide Representatives. She completed the formatting and layout of the book, helped create the figures in this book and many of my other books, and proofread this book. We are extremely grateful. We cannot thank her enough.

We thank from the bottom of our hearts Master Francisco Quintero, one of Master Sha's Worldwide Representatives, for receiving Heaven's guidance on the design of the book cover. He has contributed greatly to the mission with his spiritual channels, his teaching abilities, and more. We cannot thank him enough.

We thank from the bottom of our hearts Master Henderson Ong, also one of Master Sha's Worldwide Representatives, for designing the cover and creating many of the figures in this book. He has contributed greatly to the mission with his artistic abilities and more. We cannot thank him enough.

We thank from the bottom of our hearts Master Sha's assistant, Master Cynthia Marie Deveraux, one of Master Sha's Worldwide Representatives and one of his two Lineage Holders. She has made one of the greatest contributions to the mission. We cannot thank her enough.

We thank from the bottom of our hearts Master Marilyn Tam, the business leader of Master Sha's organization and one of Master Sha's Worldwide Representatives. Her leadership has made one of the greatest contributions to the publishing of this book and to the mission in general. We cannot thank her enough.

We thank from the bottom of our hearts Master Maya Mackie, one of the top spiritual and business leaders of the mission, one of Master Sha's Worldwide Representatives, and one of his two Lineage Holders. Her love, support, and leadership have made one of the greatest contributions to the publishing of this book and to the mission in general. We cannot thank her enough.

We thank from the bottom of our hearts all of Master Sha's business team leaders and members for their great contributions and unconditional service to the mission. We are deeply grateful. We cannot thank them enough.

We thank from the bottom of our hearts all of Master Sha's Worldwide Representatives. They are Master Teachers, servants of humanity, and servants, vehicles, and channels of the Divine. They have made incredible contributions to the mission. We thank them all deeply. We cannot thank them enough.

We thank from the bottom of our hearts the nearly seven thousand Divine Healing Hands Practitioners and Tao Hands Practitioners worldwide for their great healing service to humanity and all souls. We are deeply touched and moved. They have responded to the divine calling to serve. We deeply thank them all.

We thank from the bottom of our hearts the Tao Calligraphy Practitioners, Soul Teachers and Practitioners, Tao Song and Tao Dance Practitioners, and Soul Operation Master Practitioners worldwide for their great contributions to the mission. We are deeply touched and moved. We cannot thank them enough.

We thank from the bottom of our hearts all of Master Sha's students worldwide for their great contributions to the mission to serve humanity and all souls. We cannot thank them enough.

Master Sha thanks from the bottom of his heart his family, including his wife, his parents, his children, his brother and sisters, and more. They have all loved and supported him unconditionally. Master Sha cannot thank them enough.

Dr. Rulin Xiu thanks from the bottom of her heart her spiritual father and teacher, the creator of Tao Science, Dr. and Master Zhi Gang Sha. She cannot thank him enough for having chosen her to support him. She cannot thank him enough for having blessed and empowered her with wisdom, practices, power, connection, and big and powerful teams both in Heaven and on Mother Earth, to support her in the research of Tao Science. She wants to express her greatest gratitude for his boundless love, forgiveness, and compassion for her.

Dr. Rulin Xiu thanks from the bottom of her heart her thesis advisor and mentor, Mary K. Gaillard, and many other physicists. They have taught her and worked with her on the Grand Unified Theory. Their teaching and collaboration have been invaluable to her and to this book.

Dr. Rulin Xiu thanks from the bottom of her heart her high school Chinese teacher, Fan Xi Lin, for her love, caring, teaching, trust, and belief in her.

Dr. Rulin Xiu thanks from the bottom of her heart Soul Mind Body Science System team leaders, Kris Young, Master Janet Potts, and Marsha Valutis for their unconditional love, support, and service.

Dr. Rulin Xiu thanks from the bottom of her heart all team members in Heaven and on Mother Earth for their love, protection, support, and service to create and spread Tao Science.

Dr. Rulin Xiu thanks from the bottom of her heart Hua Feng Tian, who has saved her life.

Dr. Rulin Xiu thanks from the bottom of her heart her family, including her mother, father, sister, brother, brother-in-law, and sister-in-law for their unconditional love, trust, belief, support, and sacrifice for her.

Dr. Rulin Xiu also thanks her brother-in-law Bin He and sister Ruhong for providing a beautiful room, food, clothes, a car, and financial support for writing this book.

Dr. Rulin Xiu thanks from the bottom of her heart her native land China for giving birth to profound wisdom for humanity.

Dr. Rulin Xiu thanks from the bottom of her heart the Aloha state of Hawaii for bringing her to spiritual awakening.

We thank from the bottom of our hearts all countries for each one's special and unique wisdom. We thank from the bottom of our hearts Mother Earth and countless planets, stars, galaxies, and universes for their wisdom, nourishment, love, and beauty.

May this book serve humanity and Mother Earth by helping them pass through this difficult time in this historic period. May this book serve humanity with Tao wisdom, enlightenment, healing, rejuvenation, freedom, bliss, longevity, and immortality.

May this book serve the unification of medicine, science, spirituality, relationships, finances, and every aspect of life as One.

May this book bring love, peace, and harmony to humanity, Mother Earth, and all souls in countless planets, stars, galaxies, and universes.

May this book serve the Love Peace Harmony World Family and the Love Peace Harmony Universal Family.

May this book serve your Tao journey and the Tao journey of humanity.

We are extremely honored to be servants of you, humanity, and all souls.

Love you. Love you. Love you.
Thank you. Thank you. Thank you.

I love my heart and soul
I love all humanity
Join hearts and souls together
Love, peace and harmony
Love, peace and harmony

Bibliography

Becker, Katrin, Becker, Melanie, and Schwarz, John. *String Theory and M-Theory: A Modern Introduction.* Cambridge: University Press, 2007.

Dine, Michael. *Supersymmetry and String Theory: Beyond the Standard Model.* Cambridge: University Press, 2007.

Green, Michael, Schwarz, John H., and Edward Witten. *Superstring Theory.* Cambridge: University Press, Vol. 1: Introduction, 1987.

Green, Michael, Schwarz, John H., and Edward Witten. Cambridge: University Press, Vol. 2: *Loop Amplitudes, Anomalies and Phenomenology,* 1987.

Polchinski, Joseph. *String Theory.* Cambridge: University Press, Vol. 1: *An Introduction to the Bosonic String,* 1998.

Polchinski, Joseph. *String Theory.* Cambridge: University Press, Vol. 2: *Superstring Theory and Beyond,* 1998.

Dr. and Master Zhi Gang Sha and Dr. Rulin Xiu. *Soul Mind Body Science System: Grand Unification Theory and Practice for Healing, Rejuvenation, Longevity, and Immortality.* (Dallas/Toronto: BenBella Books/Heaven's Library Publication Corp., 2014).

Dr. Zhi Gang Sha and Dr. Rulin Xiu. "Explanation of Large-Scale Anisotropy and Anomalous Alignment from Universal Wave Function Interpretation of String Theory," to be published in 2017.

Dr. Zhi Gang Sha and Dr. Rulin Xiu. "Inflation Scheme Derived from Universal Wave Function Interpretation of String Theory," *Journal of Astrophysics & Aerospace Technology*, accepted in June 2017 for publishing.

Dr. Zhi Gang Sha and Dr. Rulin Xiu. "Dark Energy and Estimate of Cosmological Constant from String Theory," *Journal of Astrophysics & Aerospace Technology*, accepted in March 2017 for publishing.

Dr. Zhi Gang Sha and Dr. Rulin Xiu. "Space, Time, and the Creation of Universe," *Philosophy Study*, Vol. 7, No. 2, 66-75, April 2017.

Dr. Zhi Gang Sha and Dr. Rulin Xiu. "Soul Mind Body Science and Parapsychology," *Watkins' Mind Body Spirit Magazine*, November 2015.

Dr. Zhi Gang Sha and Dr. Rulin Xiu. "Can Spirit, Heart, Mind, and Consciousness Be Defined Scientifically?," *Watkins' Mind Body Spirit Magazine*, May 2015.

Selected Other Books by Dr. and Master Sha

Soul Mind Body Medicine: A Complete Soul Healing System for Optimum Health and Vitality. New World Library, 2006.

Soul Wisdom: Practical Soul Treasures to Transform Your Life. Heaven's Library Publication Corp./Atria Books, 2008.

Soul Communication: Opening Your Spiritual Channels for Success and Fulfillment. Heaven's Library Publication Corp./Atria Books, 2008.

The Power of Soul: The Way to Heal, Rejuvenate, Transform, and Enlighten All Life. Heaven's Library Publication Corp./Atria Books, 2009.

Divine Soul Mind Body Healing and Transmission System: The Divine Way to Heal You, Humanity, Mother Earth, and All Universes. Heaven's Library Publication Corp./Atria Books, 2009.

Soul Healing Miracles: Ancient and New Sacred Wisdom, Knowledge, and Practical Techniques for Healing the Spiritual, Mental, Emotional, and Physical Bodies. Heaven's Library Publication Corp./BenBella Books, 2013.

Soul Mind Body Science System: Grand Unification Theory and Practice for Healing, Rejuvenation, Longevity, and Immortality. Heaven's Library Publication Corp./BenBella Books, 2014.

About Master Sha

Gladstone, William, *Dr. and Master Sha: Miracle Soul Healer.* BenBella Books, 2014

Made in United States
North Haven, CT
10 November 2026

22091355R00157